Analysis on Manifolds

A Modern Approach to Mathematics & Physics

Analysis on Manifolds

A Modern Approach to Mathematics & Physics

V.G. Gupta

Oxford Book Company

Jaipur ▪ New Delhi

Published By

Oxford Book Company

Regd. Off. "Bony Residency", Gate No. 2
Opp. Tilak Public School, Vishweswariya Nagar,
Gopalpura Road, Jaipur - 302018, India
Ph: 0141 - 2761280, Telefax: 2761381
email - oxfordbookjpr@gmail.com
website - oxfordbookcompany.com

Delhi Office: 102, 1st Floor, Satyam House,
4327/3, Ansari Road, Darya Ganj,
New Delhi - 110002, India
Ph: 011-45652440

Distributed By

KEYA INTERNATIONAL PRESS LLP
E-mail: keyainternationalpressllp@gmail.com

Published in India, 2024

ISBN : 9788189473433

Price: ₹ 1850.00

Typeset by Shivangi Computers, Jaipur

Printed and bound in India by Thomson Press Ltd.,
New Delhi

Dedicated to our Respected Teacher

Late Professer N.D. Gautam
We pray the God to keep his soul in heaven
V.G. Gupta

Acknowledgements

I would like to express my sense of deep and sincere gratitude to Late Professor N.D. Gautam of University of Rajasthan, Jaipur, who virtually led me on the path of this world of useful and geometrical mathematical study on manifolds. I also extends my gratitude to Hon'ble vice chancellor Dr. N.K. Jain University of Rajasthan, Jaipur for blessing and encouraging to publish such a comprehensive book on Mathematics. I will be failing in my duty if I do not acknowledge my debt to Prof. M.C. Gupta, Prof. J.L. Bansal, Prof. A.N. Goyal, Prof. S.D. Mishra all of University of Rajasthan, Jaipur and Prof. K.C. Gupta, Prof. K.C. Jain of M.N.I.T. Jaipur, Dr. Rajiv Lochan ICFAI University, Jaipur for encouragement and to Prof. N.C. Jain, Head, Department of Mathematics University of Rajasthan, Jaipur for providing necessary facility in the department. My last words of gratitude must go to my parents Late Shri Chiranji Lal Gupta & Smt. Chandrakala Gupta and to my wife Indra Rajvanshi, Son Tanuj and Daughter Tanushree for extending cooperation throughout the working process of this book in varied invisible ways. We also highly appreciate the all around cooperation and support of Publisher for Publishing this book with patience, care and interest.

V.G. Gupta

Preface

The aim of writing this book is to give at the most basic level an introduction to the basic concepts which are used in differential geometry, differential equations and theoretical physics. The theory of differentiable manifolds and Lie groups is a basis of the study of above mentioned subjects and the recent development in various branches of mathematics shows that this theory is one of the cornerstones of the edifice of modern mathematics. In differential geometry we define the additional structure on the differentiable manifold namely vector field, a Riemannian metric, a 2-form, a spray, etc., and studies properties related with these objects. In differential equations, we are interested in the studies of vector fields and their integral curves, singular points, etc. It is therefore to be necessary that a certain number of concepts are essential for the study of the subjects mentioned above and so it is worthwhile to collect them together so that more advanced exposition can be given. The local coordinates are useful for computations and are also useful when integrating differential forms the study of which is necessary when one goes to higher dimensions and to study the Hamiltonian Mechanics. It is profitable to deal with infinite dimensional manifolds modelled on a Banach space.

The subject matter given in this book covers many branches of mathematics, a knowledge of which has now become essential for a modern mathematical education. We hope that a reader who has mastered this material will be able to do independent study in their field of interest among different branches of mathematics and theoretical physics. It is assumed that a knowledge of algebra, point set topology and some amount of basic analysis is essential for the study of the book.

The basic material (topological space, groups, vector spaces. calculus on Banach space) which are indispensable for the rigorous understanding of the book are gathered in Chapter 1.

The present manuscript is a organized version of the lectures delivered to the research scholars working in this field.

V.G. Gupta

Contents

1

Introduction

In this chapter, we gather the preliminary material necessary for this book. We explain the basic topics from general topology, group theory & vector space from Algebra and the differential calculus of functions of several variables. This book is written under the assumption that the reader is quite familiar with the material gathered in this chapter. We have quoted reference books for each section of this chapter. Those who are not quite familiar with the material collected here are recommended to acquaint themselves with it through those books first, in order to proceed efficiently through this book afterwards.

We shall use, without explanation, the notation ($\in, \cup, \cap, \ldots$) commonly used in set theory. **R and A** denotes the set of all real numbers and **C** the set of all complex numbers, **H** for the set of quaterion and **K** for **R, A, C** and **H**. For the numbering of the coordinates, we shall use superscripts on letters, such as $x^1, x^2, \ldots, x^n$ .That is, x^i does not mean the ith power of x, but means rather the ith coordinate. For powers of x, we shall always put parentheses around x and write $(x)^i$.

1(a). General Topology

(1.1) *Definitions: Topological Spaces*

Let *X* be a set and τ be a family of subsets of X satisfying

(τ 1.1) X itself and the empty set ϕ both belongs to τ.

(τ 1.2) If O_1, O_2 belongs to τ then $O_1 \cap O_2 \in \tau$.

(τ 1.3) If $\{O_\alpha : \alpha \in \Lambda\}$ is an arbitrary family of members of τ then the union of these members belongs to τ.

Then τ is said to be a topology defined in X and the subsets of X belonging to τ are called the open sets of this topology. A set X with a topology τ is called a topological space and is denoted by (X, τ).

A subset N of a topological space (X, τ) is called a neighbourhood of a point $x \in X$ if there exists an open set U of X such that $x \in U \subset N$.

Let N(x) denote the set of all neighborhoods of x. For the families N(x) ($x \in X$) of subsets of X, the following holds:

(N1.0) For each $x \in X$, N(x) is not empty.

(N1.1) If $U \in N(x)$, then $x \in U$.

(N1.2) If $U_1, U_2 \in N(x)$, then $U_1 \cap U_2 \in N(x)$.

(N1.3) If $U \in N(x)$ and $U \subset V$, then $V \in N(x)$.

(N1.4) If $U \in N(x)$, then there is a $V \in N(x)$, such that $V \subset U$ and $V \in N(y)$ for each $y \in V$.

Now let X be a set and suppose for each point x of a set X, a family N (x) of subsets of X is defined and satisfies the conditions (N.1.0) to (N.1.4), then a subset O of X will be called an open set of X if either O is the empty set or if O is nonempty then for an arbitrary x in O there is a $U \in N(x)$ such that $U \subset O$. If we let τ be the family of open subsets of X, then from (N.1.4) we have $N(x) \subset \tau$. Moreover it is easy to show that τ satisfies the conditions (τ.1.1), (τ.1.2), and (τ.1.3). Hence if for each point x of X a family N (x) of subsets of X is defined and if they satisfy conditions (N.1.0) to (N.1.4), then X becomes a topological space and the sets belonging to N(x) are the neighborhoods of x. We then call N(x) the fundamental neighborhood system of x or a local base at x in the topological space X.

First countable spaces

A topological space (X, τ) is said to be first countable space if each point of X possesses a countable local base.

Base for a topology

Leh (X, τ) be a topological space. A sub collection B of τ is said to form a base for τ iff for each U in τ and each x in U there is a member O of B such that $x \in O \subset U$.

Second countable spaces

A topological space (X, τ) to be second countable iff there exist a countable base for τ.

A topological space X is called a Hausdorff space, if for two arbitrary points x and y in X, $x \neq y$, there exist a neighborhood U of x and a neighborhood V of y such that $U \cap V = \varnothing$.

Let X, Y be topological spaces with fundamental neighborhood systems $N(x)$ $(x \in X)$, $N(y)$ $(y \in Y)$, respectively. For an arbitrary point (x,y) in the direct product set $X \times Y$ (the set formed by the pairs (x, y), where x is an element of X and y is an element of Y is called the direct product set of X and Y and is denoted by $X \times Y$), let $N(x, y) = \{ U \times V \mid U \in N(x), V \in N(y)\}$. Then the family $N(x, y)$ of subsets satisfies (N 1.0) to (N 1.4). Hence a topology is defined on $X \times Y$. The topological space thus defined is called the product space of the topological spaces X and Y.

Limit point, Closed set, Closure of a set

A subset F of a topological space X is said to be closed if its complement $X \sim F$is open.

Let (x, τ) be a topological space, and let A be a subset of X. The point $x \in X$ is said to be a limit point of A if every τ-neighbourhood N of x contains at least one point of A other then x, i.e.

$$(N \sim \{x\}) \cap A \neq \phi$$

The set of all limit points of a subset A of X is denoted by A' called the derived set of A. It can be proved that $A' \subset A$ it A is τ-closed. The set A is τ-closed if and only if $A' \subset A$ i.e., it contains all its limit points.

Let (x, τ) be a topological space, and let $\overline{A}$ be a subset of X. Then the smallest τ-closed subset of X containing A is called the τ-closure of A, and in denoted by A. In otherwords we can say that if F is a τ-closed set and F ⊃ A, then F ⊃ $\overline{A}$.

In a topological space (x, τ) we have, if A be a subset of X, then

$$\overline{A} = A \cup A'$$

Subspace

Let Y be a subset of the topological space X and let $\tau_Y = \{O \cap Y \mid O \in \tau\}$. The family τ_Y can easily be shown to satisfy conditions (τ1.1), (τ1.2) and (τ1.3) and thus a topology is defined in Y. The topological space Y thus defined is called a topological subspace of X.

Compact set

A family $\{O_\alpha : \alpha \in \Lambda\}$ of open subsets of the topological space X satisfying property $X = \cup \{O_\alpha : \alpha \in \Lambda\}$, is called an open covering of X.

A topological space X is said to be compact if for an aribtrary open covering $\{O_\alpha : \alpha \in \Lambda\}$ of X, one can choose a finite number of indices $\alpha_1, \ldots, \alpha_m \in \Lambda$ such that $X = O_{\alpha_1} \cup \ldots \cup O_{\alpha m}$

If a subset Y of the topological space X, considered as a subspace of X, is compact, then Y in called a compact subset of X. A compact subspace of a Hausdorff space is closed. A closed subset of a compact space is compact.

(1.2) *Continuous maps*

Let X and Y be sets. A correspondence f which associates to each element x of X an element f(x) of Y is called a mapping or a map from X into Y. If, for any two distinct elements x and y of X, the corresponding elements f(x) and f(y) are distinct, i.e., $x \neq y \Rightarrow f(x) \neq f(y)$ or $f(x) = f(y) \Rightarrow x = y$, then f is called a one-to-one map or an injection from X into Y. We also say that f is an injective. If, for an arbitrary element y of Y, there is always an element x of X such that f(x) = y, then f is called an onto map or a surjection. We also say that f is surjective. When f is an injection and a surjection, we call f a bijection. We also say that f is bijective function.

A mapping f from a topological space X to a topological space Y is called a continuous map if for an arbitrary open set O of Y, the inverse image of O by f, i.e. $f^{-1}(O) = \{x \in X \mid f(x) \in O\}$, is always an open set of X.

If f is a bijection from X onto Y, and if f and its inverse f^{-1} [i.e., the map from Y to X defined by $f^{-1}(y) = x$ if $y = f(x)$] are both continuous maps, then we call f a homeomorphic map or simply a homeomorphism from X onto Y. If there is a homeomorphism from X onto Y, then we say that X and Y are homeomorphic.

Also the continuity of $f: X \to Y$ is equivalent to the following condition: for an arbitrary x in X and an arbitrary neighborhood V of f(x), there is a neighborhood U of x such that $f(U) \subset V$.

We also have if f is a continuous map from X to Y, and g is a continuous map from Y to Z, then the composition gof from X to Z (defined by $(gof)(x) = g(f(x))$ for $x \in X$) is continuous. If f is a continuous map from X into Y and A is a subset of X, then the restriction $f \upharpoonright A$ (defined by $(f \upharpoonright A)(x) = f(x)$ for $x \in A$) is continuous map from A into Y.

(1.3) *Connected sets*

Let X be a topological space. Then X is said to be disconnected if there are two open sets O_i (i = 1,2) of X satisfying $O_i \neq \phi$ (i = 1,2), $O_1 \cap O_2 = \phi$ and $X = O_1 \cup O_2$. If X is not disconnected then X is said to be connected, that is X is connected if the following condition is satisfied: if O_1, O_2 are two open subsets of X satisfying

$O_1 \cap O_2 = \phi$ and $X = O_1 \cup O_2$, then either O_1 or O_2 is the empty set. Let Y be a subset of the topological space X. If Y, considered as a topological subspace of X, is connected, then we call Y a connected subset of X. A continuous image of a connected space is connected.

If $Y_\alpha (\alpha \in \Lambda)$ are connected subsets of X, and if $\bigcap_{\alpha \in \Lambda} Y_\alpha$ is not empty, then the union is also connected. In fact, suppose Y is not connected. Then there are open sets U, V of X satisfying $U \cap Y \neq \phi$, $V \cap Y \neq \phi$, $U \cap V \cap Y = \phi$ and $Y = (U \cap Y) \cup (V \cap Y)$. Let $U \cap Y_\alpha = U_\alpha$, $V \cap Y_\alpha = V_\alpha$. Then U_α, V_α are open sets of Y_α and $\alpha \in \Lambda''$ hold. However, since Y_α is connected, $U_\alpha = \phi$ or $V_\alpha = \phi$. Now let $\Lambda' = \{\alpha \in \Lambda: U_\alpha = \phi\}$ and $\Lambda'' = \{\alpha \in \Lambda: V_\alpha = \phi\}$. Then $\Lambda = \Lambda' \cup \Lambda''$, and $Y_\alpha = V_\alpha$ if $\alpha \in \Lambda'$ and $Y_\alpha = U_\alpha$ if.

Hence $\bigcap_{\alpha \in \Lambda} Y_\alpha = \left(\bigcup_{\alpha \in \Lambda'} Y_\alpha\right) \cap \left(\bigcap_{\alpha \in \Lambda''} Y_\alpha\right) \subset (Y \cap V) \cap (Y \cap U)$.

However, $(Y \cap V) \cap (Y \cap U) = \phi$, so we have $\bigcap_{\alpha \in \Lambda} Y_\alpha = \phi$, contrary to our hypothesis that $\bigcap_{\alpha \in \Lambda} Y_\alpha$ is nonempty. Hence Y must be connected.

In particular, the union of all connected subsets of X containing a point x of X is a connected subset of X, this is called the connected component of X containing x. This is the maximum connected subset containing x. Let C be the connected component of X containing x. If y is in C, it is easy to see that C is also the connected component of X containing x, then we call C a connected component of X. If C and C' are connected compnents of X, and if $C \neq C'$, then $C \cap C' = \phi$.

A topological space X is said to be locally connected if each point has a fundamental neighborhood system consisting of connected neighborhoods. If X is locally connected, then each connected component of X is an open set of X.

A continuous map c from a closed interval [a,b] of the real line to a topological space X is called a continuous curve in X. For two points x and y of X, a continuous curve c such that c (a) = x and c (b) = y is called a continuous curve connecting x and y. x = c(a) is called the starting point of c and y = c(b) is called the terminal point of c.

If for any two points x and y in a topological space X, there is always a continuous curve in X connecting x and y, then X is said to be arcwise connected. If X is arcwise connected, then X is connected.

(1.4) *Metric spaces*

Let X be a set. Then a real valued function $d: X \times X \to \underline{R}$ is called a metric on X if for any pair of points x, y of X, there is a real number d(x, y) satisfying the following conditions:

(M 1.1) $d(x, y) > 0, \ \forall \ x, y \in X$
(M 1.2) $d(x, y) = 0$ iff $x = y$
(M 1.3) $d(x, y) = d(y, x)$,
(M 1.4) $d(x, y) \leq d(x, z) + d(z, y)$ for all $x, y, z \in X$.

A set X with a metric d is called a metric space. For a point x in a metric space X, and for $r > 0$, let $B(x; r) = \{y \in X \mid d(x,y) < r\}$, called the open ball in X, $B[x; r] = \{y \in X \mid d(x,y) \leq r\}$ is called the closed ball in X and $S(x; r) = \{y \in X \mid d(x,y) = r\}$ is called the sphere in X . If we let $N(x) = \{B(x;r) \mid 0 < r < 1\}$. then N(x) satisfies (N 1.0) to (N 1.4) . Thus a topology is defined in the metric space X. Hereafter, we shall always assign this topology to a metric space X and consider X as a topological space.

Let $\{x_n \mid n = 1, 2...\}$ be sequence of points in X. If, for a given $r > 0$, there is a positive integer N such that $d(x_n, x) < r$ holds for $n > N$, then we say that the sequence $\{x_n\}$ converges to a point x in X. We say that a sequence $\{x_n\}$ of points in X is a Cauchy sequence if, for a given $r > 0$, there is a positive integer N such that $d(x_n, x_m) < r$ holds for arbitrary m, n satsfying $m, n > N$. A sequence converging to a point in X is always a Cauchy sequence. If an arbitrary Cauchy sequence in X always converges to a point in X, then we say that the metric space X is complete. Omission of a point a from the real line yields the incomplete space $R \sim \{a\}$. Also by the omission of all irrational numbers from the real line we have the rational line Q, which is incomplete. An open interval (a,b) with the metric induced from R is another incomplete metric space. Let $X = (0, 1]$, with the usual metric defined by $d(x, y) = |x - y|$ and the sequence (x_n), where $x_n = 1/n$, n = 1,2,.... This is a cauchy sequence, but it does not converge, because the point 0 to which it wants to converge is not a point of X.

We can make use of the metrics to define the continuity of a map f from a metric space X to a metric space Y. A map $f : X \rightarrow Y$ is continuous at a point $x \in X$ if for every $\in > 0$ there exists $r > 0$ such that $d(f(x), f(a)) < \in$ whenever $d(x, a) < r$.

Let X and Y be metric space with metrices d and d' respectively. We say that a map f from X to Y is uniformly continuous if for a given $\varepsilon > 0$ there is a $\delta > 0$ such that $d'(f(x), f(y)) < \varepsilon$ whenever $d(x, y) < \delta$ for any two points x, y of X. If X is compact, then a continuous map from X to Y is uniformly continuous.

1(b). Preliminary from Algebra

(1.5) *Group Theory*

Definition Group: A group $(G, *)$ is an algebric structure consisting of a non-empty set G equipped with one binary operation $* : G \times G \rightarrow G$ associating each elements $(a, b) \in G \times G$ to a unique element $a * b \in G$ and satisfying the following axioms.

(1) $(a*b)*c = a*(b*c)$ for all $a, b, c \in G$ (Associativity)

(2) There exists an elements e in G such that $a*e = e*a = a$ for all $a \in G$ (Existence of identity)

(3) For each $a \in G$ there exist an element $b \in G$ such that $a*b = e = b*a$ (Existence of Inverse).

Abelian or Commutative group:- A Group $(G, *)$ is called abelian or Commutative if $a*b = b*a$ for all $a, b \in G$

(1.6) *Theorem*

Let $(G, *)$ be a group then:

(1) The identity element in $(G, *)$ is unique.

(2) The inverse of an element in $(G, *)$ is unique.

(3) $(a*b)^{-1} = b^{-1} * a^{-1}$ for all $a, b \in G$

(4) The equations $a*x=b$ and $y*a=b$ have a unique solutions in G where $a,b \in G$.

Proof :

(1) Let e_1 and e_2 are two identity elements in G, since e_1 is the left identity in G, $e_1*e_2=e_2$ for $e_2 \in G$, since e_2 is the right identity in G $e_1*e_2=e_1$ for $e_1 \in G$, hence $e_1=e_2$.

(2) Let a be any element in G, let b and c are inverses of a in G so $a*b=b*a=e$ and $a*c=c*a=e$ consider $b*a*c=(b*a)*c=e*c=c$ also $b*a*c=b*(a*c)=b*e=b \Rightarrow b=c$.

(3) Let $a,b \in G \Rightarrow a^{-1},b^{-1} \in G \Rightarrow b^{-1}*a^{-1} \in G$. Let $c=b^{-1}*a^{-1}$

Now $(a*b)*c=(a*b)*(b^{-1}*a^{-1})=a*(b*b^{-1})*a^{-1}$

$=a*e*a^{-1}=a*a^{-1}=e$,

and $c*(a*b)=(b^{-1}*a^{-1})*(a*b)=b^{-1}*(a^{-1}*a)*b$

$=b^{-1}*e*b=b^{-1}*b=e$

Hence $(a*b)^{-1}=c=b^{-1}*a^{-1}$

(4) Let $a,b \in G \Rightarrow a^{-1},b^{-1} \in G \Rightarrow a^{-1}*b \in G$

Put $x=a^{-1}*b$ in $a*x=b$ we have $b=a*(a^{-1}*b)=(a*a^{-1})*b=e*b=b$

which shows $x=a^{-1}*b$ is a solution of $a*x=b$ in G. To prove uniqueness, let x_1 and x_2 be two different solution of $a*x=b$ in G, then $a*x_1=b,\ a*x_2=b$

Thus, $a*x_1=a*x_2$

$\Rightarrow x_1=x_2$ by left cancellation law. Similarly $y=b*a^{-1}$ is a unique solution of $y*a=b$ in G.

(1.7) *Examples*

(1) (Z,+), (Q,+), (R,+) and (C,+) are all groups with respect to addition as binary operation, where Z is the set of integers, Q is the set of rational numbers, R is the set of real numbers and C is the set of complex numbers.

(2) $\langle Q^*=Q\sim\{0\},\ \cdot\rangle$, $\langle R^*=R\sim\{0\},\ \cdot\rangle$, $\langle C^*=C\sim\{0\},\ \cdot\rangle$ are all groups with respect to multiplication as binary operation.

(3) Z is not a group with respect to multiplication.

(4) $R^+ = \{x \in R \mid x > 0\}$ is the group with respect to multiplication.

(5) $R^n =$ the set of all orderd n-tuples of real numbers is a group with respect to addition operation : if $x = (x_1, \ldots, x_n)$ and $y = (y_1, \ldots, y_n)$ then $x + y = (x_1, \ldots, x_n) + (y_1, \ldots, y_n) = ((x_1 + y_1), \ldots, (x_n + y_n))$ there exists $0 = (0, \ldots, 0)$ and $-x = (-x_1, \ldots, -x_n)$ in R^n satisfies the condition of identity and inverse element.

(6) Let GL(n,R) be the set of all $n \times n$ invertible real matrices, i.e., $GL(n, R) = \{A^{n \times n} \mid \det A \neq 0\}$. Then GL(n, R) is a group with respect to multiplication of matrices. The identity element is unit matrix I_n and inverse of A is $A^{-1} = \dfrac{\mathrm{Adj}(A)}{|A|}$.

The group GL(n, R) is called the general linear group of degree n over R. Similarly

$GL(n, C) = \{A^{n \times n} \mid \det A \neq 0\}$ is also a group for multiplication, called the general linear group of degree n over C.

(7) Let $\underline{n} = \{1, \ldots, n\}$, then the set S_n of all permutation of the set $\underline{n}$ is a group of order n! with respect to composition of permutation as binary operation, it is called the symmetric group of degree n. The symmetric group of degree 3, i.e., $(S_3, 0)$ is an example of a non-abelian group and it is the group of all symmetries of an equilateral triangle in a plane.

(8) Let G=SO(2), the group of rotation in the plane, i.e.,

$$G = \left\{ \begin{pmatrix} \cos\theta & -\sin\theta \\ \sin\theta & \cos\theta \end{pmatrix} : 0 \leq \theta < 2\pi \right\}$$

(where θ denote the angle of rotation) is a group for multiplication, we can identify G with the unit circle $S^1 = \{(\cos\theta, \sin\theta) : 0 \leq \theta < 2\pi\}$ in R^2.

(9) The set $O(2) = \{A \in GL(2,R): A^T A = I\}$ is a group for multiplication operation called the orthogonal group.

(1.8) *Sub group*

Definition: A non-void subset H of a group G is a subgroup of G if H itself is a group for the induced binary operation defined in G. Every group G has two subgroups namely G and {e}, these two subgroups are known as trivial subgroup of G or an improper subgroups of G, subgroups of G other than these two are called proper subgroups of G.

(1.9) *Theorem*

The necessary and sufficient condition that a non-void subset H of a group $(G,*)$ to be a subgroup is that $a \in H, b \in H \Rightarrow a*b^{-1} \in H$.

Examples:

(1) The additive group of complex numbers C have the following chain of subgroups $Z \subseteq Q \subseteq R \subseteq C$.

(2) The multiplicative group of complex number C^* have the following chain of subgroups $Z^* \subseteq Q^* \subseteq R^* \subseteq C^*$ where $Z^* = \{-1,1\}$.

(3) The set H of $n \times n$ real matrices with determinant one is a subgroup of GL(n, R). The set H is a subset of GL(n, R) because any matrix with determinant one is invertible. This group is called the special linear group over the reals and is denoted SL(n, R).

(4) Let $H \subseteq GL(2,R)$ given by

$$H = \left\{ \begin{pmatrix} 1 & 0 \\ 0 & 1 \end{pmatrix}, \begin{pmatrix} -1 & 0 \\ 0 & 1 \end{pmatrix}, \begin{pmatrix} 1 & 0 \\ 0 & -1 \end{pmatrix}, \begin{pmatrix} -1 & 0 \\ 0 & -1 \end{pmatrix} \right\},$$ then H is an abelian subgroup of GL(2,R), every non-identity element has order 2.

(5) The center of a group G is the set $\{x \in G \mid gx = xg, \text{ for all } g \in G\}$, it is also a subgroup of G.

(6) Let a be any element of G. The normalizer of a in G is the set $N(a) = \{x \in G \mid ax = xa\}$ which is also a subgroup of G.

(1.10) *Definition*

A function on groups which preserve the operations is called homomorphism.

Let $(G,*)$ and $(G',\circ)$ are two groups, a function $f:G \to G'$ is called a homomorphism if for all $a, b \in G$ we have $f(a*b) = f(a) \circ f(b)$.

(1.11) *Theorem*

If $f:G \to G'$ be a homomorphism, then we have

(1) $f(e) = e'$, e is the identity in G and e' is the identity in G'.

(2) $f(a^{-1}) = (f(a))^{-1}$, $a \in G$.

Proof:

(1) Let e and e' be identies in G and G' respectively, let $a \in G$ then

$f(a) \cdot e' = f(a) = f(ae) = f(a)f(e)$ hence by left cancellation law $e' = f(e)$.

(2) Now $e' = f(e) = f(a\,a^{-1}) = f(a) \cdot f(a^{-1})$ shows

$$f(a^{-1}) = (f(a))^{-1}$$

A homomorphism $f:G \to G'$ is called an epimophism if it is surjective (or onto), i.e., $f(G) = G'$. Define $f:(R,+) \to (R^2,+)$ by $f(x) = (x, x)$ $\forall x \in R$, then f is a homomorphism but it is not surjective because $f(x)$ is just the diagonal line in R^2. The function $g:(R^2, +) \to (R,+)$ defined by $g(x,y) = x$ is a surjective homomorphism. A homomorphism $f:G \to G'$ is called monomorphism if it is injective (or one-one), i.e., $f(a) = f(b) \Rightarrow a = b$. The function f defined above is injective but g is not injective. A homomorphism $f:G \to G'$ is called an isomorphism if it is both injective (or one-one) and surjective. Two groups are isomorphic means they are really the same group-even if they were defined in strikingly different manner. Let R be the addition group of real number and R^+ be the multiplicative group of all positive real numbers. Let a be any real number greater than 1, define $f:(R,+)\to(R^+,\cdot)$ by $f(x)=a^x \ \forall x\in R$, then f is a homomorphism $f(x+y)=a^{x+y}=a^x a^y=f(x)f(y)$. Also, f is injective, for, suppose $f(x)=f(y)$, this means $a^x=a^y \Rightarrow a^x\cdot a^{-y}=a^y\cdot a^{-y}\Rightarrow a^{x-y}=1$, which implies x - y = 0 or x = y. Also f is surjective, for, if y is any positive real number there exists $x=\log_a y$ such that $a^x=y$. Thus these two groups are isomorphic, not only that, but there are lots of isomorphism. An isomorphism of a group with itself is called an automorphism.

Kernal of f

If $f:G\to G'$ be a homomorphism, and e′ is the identity element of G′. Then the Kernel(f) = $\{x\in G \mid f(x)=e'\}$. It can be easily proved that Kernel of f is a subgroup of G.

Normal sub group

A subgroup N of a group G is called a normal subgroup of G if for each element g of G and each element n of N, the element gng^{-1} belongs to N. If G is a commutative, then every subgroup of G is automatically normal.

It is easy to prove that Ker(f) is a normal subgroup of G. Let G = G′ = Z and define f(n) = 2n. This is a homomorphism of Z to itself, but not an automorphism. The Kernel of this homomorphism is just $\{0\}$. The determinant is a homomorphism of GL(n, R) to $R_0=R\sim\{0\}$. The Kernel of this map is SL(n, R).

Given any group G and any element g in G, define $\Phi_g:G\to G$ by $\Phi_g(x)=gxg^{-1}$. Then Φ_g is an automorphism of G. Also the map $g\mapsto\Phi_g$ is a homomorphism of G into Aut(G) and the Kernel of this map is the center of G. The automorphism Φ_g is called an inner automorphism.

Quotient group

Let G be a group and N be a normal subgroup of G. Then the set $\frac{G}{N}=\{Na|a\in G\}$, i.e., the set of all distinct cosets of N in G is a group with respect to the operation defined in $\frac{G}{N}$ as given below Na.Nb = Nab. Thus $\left(\frac{G}{N},\cdot\right)$ is a group called the quotient group.

1(c) Field and Quaternions

(1.12) *Definition*

A field F is a non empty set equipped with two binary operations namely addition and multiplication satisfying the following axioms.

(1) (F,+) is an abelian group.

(2) $(F\sim\{0\},\cdot)$ is an abelian group.

(3) Multiplication distributes over addition a.(b + c) = $a\cdot b+a.c\quad \forall a,b,c\in F$.

(1.13) *Examples*

The set Q of rationals and the set R of reals are fields we can make R^2 into a field C (the complex numbers) as follows.

Define addition and multiplication in R^2 as follows, for each $(x_1,x_2),(y_1,y_2)\in R^2$,

$(x_1,x_2)+(y_1,y_2)=(x_1+y_1,x_2+y_2)$ and

$(x_1,x_2)\cdot(y_1,y_2)=(x_1y_1,x_2y_2)$.

Then $(R^2,+)$ is an abelian group, but for multiplication operation we have $(1,0)\cdot(0,1)=(0,0)$, Shows that product of two non-zero elements in R^2 is a zero element of R^2, so $R^2\sim\{0\}$ is not an abelian group, consequently $(R^2,+,\cdot)$ is not a field.

Now if we define the multiplication operation in R^2 as the multiplication of complex numbers, i.e., $(x_1, x_2)\cdot(y_1, y_2) = (x_1y_1 - x_2y_2, x_1y_2 + x_2y_1)$ then $(R^2, \cdot)$ satisfies all the properties for an abelian group, i.e.,(1,0) is an multiplicative identity, the inverse of $(a,b) \neq (0,0)$ is $\left(\frac{a}{a^2+b^2}, \frac{-b}{a^2+b^2}\right)$ and the distributive property is also satisfied, i.e., $(a,b)((c,d)+(e,f)) = (a,b)(c+e, d+f)$

$$= (a(c+e) - b(d+f), a(d+f) + b(c+e)) = (a,b)(c,d) + (a,b)(e,f)$$

Thus we have made R^2 into a field which we denote by C and call the field of complex numbers. We can consider R to be a subfield of C by letting $x \in R$ be $x + i0$.

Then if $x, y \in R$ we have $x + y = (x + i0) + (y + i0) = (x + y) + i0$

$x \cdot y = (x + i0)\cdot(y + i0) = (x \cdot y) + i0$ so we have taken the field R as all (x,0) in R^2 and extended the operation in R to R^2 to get a field.

Now the question arises can we extend the field R^2 to a field R^3.

(1.14) ***Theorem***

The operation on C cannot be extended to make R^3 into a field.

Proof: Take 1,i,j be basis vectors, so that any element of R^3 can be written uniquely as $a + ib + jc$ with $a, b, c \in R$. Let ij=a+ib+jc,for i,j and $a, b, c \in R$, but then

$$i(ij) = ia + i^2b + ijc \text{ so } -j = ia - b + ijc, \ (i^2 = -1),$$

$$-j = ia - b + (a + ib + jc)c = (ac - b) + i(a + bc) + jc^2$$

this implies $c^2 = -1$, which is a contradiction as $c \in R$. The main thrust of this proof is that if we insist that the product ij be in R^3 we get into trouble. May be if we had one more dimension it would work. We can define a multiplication on R^4 which satisfied conditions

(1) and (3) for a field but (2) must be replaced by (2)`, i.e., $F \sim \{0\}$ is a group under multiplication it is not an abelian group.

We take a basis 1, i, j, k for R^4 and define

	1	i	j	k
1	1	i	j	k
i	i	−1	k	−j
j	j	−k	−1	i
k	k	j	−i	−1

Thus 1 acts as identity, ij=k, ji=-k etc. This tells us how to multiply quadruples of real numbers.

The Irish mathematician Rowan Hamilton realized that it is necessary to go to four dimensions and so he invented a new number system called the quaternions ,it is denoted by for Hamilton and it is an ordered pairs of complex numbers, i.e., $q = (z_1, z_2)$, $z_1, z_2 \in C$, with addition and multiplication defined by

$$(z_1, z_2) + (w_1, w_2) = (z_1 + w_1, z_2 + w_2)$$

$$(z_1, z_2) \cdot (w_1, w_2) = (z_1 w_1 - z_2 w_2^*, z_1 w_2 + z_2 w_1^*) \text{ and}$$

$$a(z, w) = (a,0)(z, w) = (az, aw) \quad a \in R$$

It turns out that multiplication is not commutative. That is, in general , $rq \neq qr$, another representation of quaternions is in the form $q_0 + q_1 i + q_2 j + q_3 k$, $q_i \in R$ .The quaternions i,j,k satisfiy the following relations $i^2 = j^2 = k^2 = ijk = -1$ ...(1.14.1)

The addition and multiplication operations in are defined as follows

$$(q_0 + q_1 i + q_2 j + q_3 k) + (p_0 + p_1 i + p_2 j + p_3 k) =$$
$$(q_0 + p_0) + (q_1 + p_1)i + (q_2 + p_2)j + (q_3 + p_3)k$$

and $(q_0 + q_1 i + q_2 j + q_3 k) \cdot (p_0 + p_1 i + p_2 j + p_3 k) =$

$$(q_0 p_0 - q_1 p_1 - q_2 p_2 - q_3 p_3) + (q_0 p_1 + q_1 p_0 + q_2 p_3 - q_3 p_2)i$$
$$+(q_0 p_2 + q_2 p_0 + q_3 p_1 - q_1 p_3)j + (q_0 p_3 + q_3 p_0 + q_1 p_2 - q_2 p_1)k$$

given $q = q_0 + q_1 i + q_2 j + q_3 k$, q_0 is called the scalar part of q, i.e., $S(q) = q_0$ and $q_1 i + q_2 j + q_3 k$ is called the vector part of q, i.e., $V(q) = q_1 i + q_2 j + q_3 k$, the conjugate of q is $\bar{q} = q_0 - q_1 i - q_2 j - q_3 k$, the square of the norm of q is $|q|^2 = q\bar{q} = q_0^2 + q_1^2 + q_2^2 + q_3^2 \in R$ when $|q| \neq 0$, the inverse of q is $q^{-1} = \frac{1}{|q|^2}\bar{q}$,

clearly $qq^{-1} = 1 + 0i + 0j + 0k$. Note also that $\overline{qp} = \bar{p}\,\bar{q}$ and $(qp)^{-1} = p^{-1}q^{-1}$. Thus we have express quaternions in two ways firstly the complex number pair form and secondly the i,j,k form, there is a third way which is convenient for any one accustomed to vector algebra, this third way represent, quaternions as scalar vector pairs,

$q = q_0 + q_1 i + q_2 j + q_3 k \leftrightarrow (q_0, \mathbf{v})$ with $\mathbf{v} = q_1\mathbf{e}_1 + q_2\mathbf{e}_2 + q_3\mathbf{e}_3$. In this way, the quaternion algebra is expressed as

$$(q_0, \mathbf{u}) + (p_0, \mathbf{v}) = (q_0 + p_0, \mathbf{u} + \mathbf{v})$$

$$(q_0, \mathbf{u}) \cdot (p_0, \mathbf{v}) = (q_0 \cdot p_0 - \mathbf{u} \cdot \mathbf{v}, q_0 \cdot \mathbf{v} - p_0 \cdot \mathbf{u} + \mathbf{u} \times \mathbf{v})$$

$$\overline{(q_0, \mathbf{u})} = (q_0, -\mathbf{u})$$

$$|(q_0, \mathbf{u})|^2 = q_0^2 + \mathbf{u} \cdot \mathbf{u}$$

$$(q_0, \mathbf{u})^{-1} = \frac{1}{q_0^2 + \mathbf{u} \cdot \mathbf{u}}(q_0, -\mathbf{u})$$

a pure quaternion $(0, \mathbf{u})$ is one with zero scalar part and a unit quaternion has unit norm $q\bar{q} = 1$, The product of two pure quaternions encapsulates the scalar and vector products $(0, u)(0, v) = (-u \cdot v, u \times v)$.

(1.15) *Quaternions as Matrices*

Complex numbers can be represented as 2×2 matrices with real entries. Let $J=\begin{pmatrix}0 & -1\\ 1 & 0\end{pmatrix}$ and note that $J^2=-1$, then Z=aI+bJ and W=cI+dJ satisfiy the rules for addition and multiplication of complex numbers,

$$Z+W=(a+c)I+(b+d)J$$

$$ZW=(aI+bJ)(cI+dJ)=(ac-bd)I+(ad+bc)J$$

similarly, quaternions can be represented as 2×2 matrices of complex number. If

$z_1=q_0+iq_1,\ z_2=q_2+iq_3,\ w_1=r_0+ir_1,\ w_2=r_2+ir_3$ then the correspondences

$$Q_1=\begin{pmatrix}z_1 & z_2\\ -z_2^* & z_1^*\end{pmatrix}\leftrightarrow q_0+q_1i+q_2j+q_3k \quad \text{and}$$

$$Q_2=\begin{pmatrix}w_1 & w_2\\ -w_2^* & w_1^*\end{pmatrix}\leftrightarrow r_0+r_1i+r_2j+r_3k$$ are preserved by the matrix product

$Q_1Q_2\leftrightarrow(q_0+q_1i+q_2j+q_3k)(r_0+r_1i+r_2j+r_3k)$. The quaternion product is linear in the component of each factor. This allows us to exprees the quaternion operations in terms of linear operations on R^4. If we associate

$$q=q_0+q_1i+q_2j+q_3k\leftrightarrow q=\begin{pmatrix}q_0\\ q_1\\ q_2\\ q_3\end{pmatrix} \quad \text{and}$$

$$p=p_0+p_1i+p_2j+p_3k\leftrightarrow p=\begin{pmatrix}p_0\\ p_1\\ p_2\\ p_3\end{pmatrix}$$

then left multiplication by q is represented by

$$L(q)=\begin{pmatrix} q_0 & -q_1 & -q_2 & -q_3 \\ q_1 & q_0 & -q_3 & q_2 \\ q_2 & q_3 & q_0 & -q_1 \\ q_3 & -q_2 & q_1 & q_0 \end{pmatrix}$$

i.e., $qp \leftrightarrow L(q)p$. This is expressed explicitly in a component form

$$\begin{pmatrix} q_0 & -q_1 & -q_2 & -q_3 \\ q_1 & q_0 & -q_3 & q_2 \\ q_2 & q_3 & q_0 & -q_1 \\ q_3 & -q_2 & q_1 & q_0 \end{pmatrix}\begin{pmatrix} p_0 \\ p_1 \\ p_2 \\ p_3 \end{pmatrix}=\begin{pmatrix} q_0p_0-q_1p_1-q_2p_2-q_3p_3 \\ q_0p_1+q_1p_0+q_2p_3-q_3p_2 \\ q_0p_2-q_1p_3+q_2p_0+q_3p_1 \\ q_0p_3+q_1p_2-q_2p_1+q_3p_0 \end{pmatrix}$$

Right multiplication by q is represented by

$$R(q)=\begin{pmatrix} q_0 & -q_1 & -q_2 & -q_3 \\ q_1 & q_0 & q_3 & -q_2 \\ q_2 & -q_3 & q_0 & q_1 \\ q_3 & q_2 & -q_1 & q_0 \end{pmatrix}$$

i.e., $pq \leftrightarrow R(q)p$. These are the representation of the Left and Right translation L_q and R_q on the Lie group $S^3 \cong H_1$ defined in chapter 12. Since $L_pL_q = L_{pq}$, it follows that $L(p)L(q) = L(pq)$. There the analogous result for Right multiplication $R(p)R(q) = R(qp)$, $L(p)$ and $R(p)$ have the following properties

$$L(q)p = -L(p)q \quad \text{if } q^Tp = 0 \text{ and } R(q)p = -R(p)q \text{ if } q^Tp = 0$$

$$L(\bar{q}) = L(q)^T,\ L(\bar{q})\,L(q) = (q^Tq)I,\ R(\bar{q})\,R(q) = (q^Tq)I, \text{ if } q \in H_1$$

then $R(q)$ and $L(q)$ are orthogonal matrices.

1(d) Vector Spaces:

(1.16) *Definition of Vector Spaces and Basis*

Let the symbol K stand for R,C or H. (The reader who is familiar with the notion of a "field" may consider K to be an arbitrary field.)

A set V with two binary operation namely internal operation addition denoted by '+' and external operation scalar multiplication denoted by '.' and satisfying the condition in (I) and (II) is called a vector space over K.

(I) For any two elements **a**, **b** of V an element **a** + **b** of V, called the sum of **a** and **b**, is defined, and the following rules hold:

(V 1.1) $(\mathbf{a}+\mathbf{b})+\mathbf{c}=\mathbf{a}+(\mathbf{b}+\mathbf{c})$ (Associativity)

(V 1.2) $\mathbf{a}+\mathbf{b}=\mathbf{b}+\mathbf{a}$ (Commutativity)

(V 1.3) An element **0** of V, called the zero vector, exists and for an arbitrary $\mathbf{a} \in V$ satisfies $\mathbf{a}+\mathbf{0}=\mathbf{a}=\mathbf{0}+\mathbf{a}$ (existence of identity)

(V1.4) For an arbitrary $\mathbf{a} \in V$, there is an element $-\mathbf{a}$ of V, called the additive inverse element of **a**, which satisfies. $\mathbf{a}+(-\mathbf{a})=\mathbf{0}$ (existence of inverse)

(II) For an element **a** of V and an element λ of K, an element $\lambda\mathbf{a}$ of V, called λ times **a**, is defined. $\lambda\mathbf{a}$ is called the scalar multiple of **a** by λ. For the addition in (I) and the scalar multiplication in (II), the following rules must hold:

(V2.1) $(\lambda\mu)\mathbf{a}=\lambda(\mu\mathbf{a}),\ (\lambda,\mu \in K)$

(V2.2) $1\mathbf{a}=1$

(V2.3) $\lambda(\mathbf{a}+\mathbf{b})=\lambda\mathbf{a}+\lambda\mathbf{b}\ \ \mathbf{a},\mathbf{b} \in V$

(V2.4) $(\lambda+\mu)\mathbf{a}=\lambda\mathbf{a}+\mu\mathbf{a}$

The elements of a vector space V over K are called vectors and the elements of K are called scalars. We also call K the coefficient field of the vector space V. In particular, if K = R we call V a real vector space, and if K = C we call V a complex vector space.

Let $\mathbf{a}_1, \ldots, \mathbf{a}_n$ be elements of V. Then $\mathbf{a}_1, \ldots, \mathbf{a}_n$ are said to be linearly independent if for n scalars $\lambda^1, \ldots, \lambda^n$, we have $\lambda^1\mathbf{a}_1 + \ldots + \lambda^n\mathbf{a}_n = \mathbf{0}$, then $\lambda^1 = \ldots = \lambda^n = 0$.

If n vectors are linearly independent in V, and n+1 or more vectors are never linearly independent, then we say that V has dimension n. When V is of dimension n, a set of n linearly independent vectors $\{\mathbf{a}_1, \ldots, \mathbf{a}_n\}$ which also span V is called a basis of V. The basis vectors will be indexed with subscripts. If $\{\mathbf{e}_1, \ldots, \mathbf{e}_n\}$ are standard basis of V, then an arbitrary vector $v = (v^1, \ldots, v^n)$ of V can be written uniquely as $v = v^i \mathbf{e}_i = v^1 \mathbf{e}_1 + \ldots + v^n \mathbf{e}_n$. Here **e** and $\boldsymbol{v}$ are denoted as $\mathbf{e} = (\mathbf{e}_1, \ldots, \mathbf{e}_n)$, $\boldsymbol{v} = (v^1, \ldots, v^n)^T$ and the representation of a vector $\boldsymbol{v}$ in the basis **e** is denoted by $\boldsymbol{v} = \mathbf{e}\, v$. The basis **e** is also be referred to as a frame. The n-tuple of numbers $(v^1, \ldots, v^n)$ are called the components (or coordinates) of $\boldsymbol{v}$ with respect the basis $(\mathbf{e}_1, \ldots, \mathbf{e}_n)$.

The distinction between a vector $\boldsymbol{v} = \mathbf{e}\, v$ and its components v relative to a basis **e** is of prime importance. The former is invariant under a change of basis and the latter of cource is not. We note that K itself is a vector space over K of dimenstion 1.

(1.17) *Linear maps*

Let **V** and **W** be vector spaces over K. A map f from **V** to **W** satisfying.

$$f(\mathbf{a} + \mathbf{b}) = f(\mathbf{a}) + f(\mathbf{b}), (\mathbf{a}, \mathbf{b} \in \mathbf{V}) \text{ and } f(\lambda \mathbf{a}) = \lambda f(\mathbf{a}), \quad (\lambda \in K, \mathbf{a} \in \mathbf{V}), \quad \ldots(1.17.1)$$

or equivalently $f(\lambda\, \mathbf{a} + \mu\, \mathbf{b}) = \lambda\, f(\mathbf{a}) + \mu\, f(\mathbf{b})$

is called a linear map and a linear transformation if **V** = **W**. Let the dimensions of **V** and **W** be m and n respectively, let f be a linear map and $\{\mathbf{a}_1, \ldots, \mathbf{a}_m\}$ and $\{\mathbf{b}_1, \ldots, \mathbf{b}_n\}$ be basis of **V** and **W** respectively. As $f(\mathbf{a}_i)$ is an element of **W**, it can be expressed uniquely as

$f(\mathbf{a}_i) = \sum_{j=1}^{n} A_{ij} b_j$, (i=1,...,m , $A_{ij} \in K$).Hence an n ×m matrix

$$A = \begin{bmatrix} A_{11} & A_{12} & \dots & A_{1n} \\ A_{21} & A_{21} & \dots & A_{2n} \\ \vdots & \vdots & & \vdots \\ A_{m1} & A_{m2} & \dots & A_{mn} \end{bmatrix}^T$$

corresponds to the linear map f. For $\mathbf{a} \in \mathbf{V}$, let $\mathbf{a} = \lambda^1 \mathbf{a}_1 + \dots + \lambda^m \mathbf{a}_m$ and $f(\mathbf{a}) = \mu^1 \mathbf{b}_1 + \dots + \mu^n \mathbf{b}_n$. Then we have

$$f(\mathbf{a}) = \lambda^1 f(\mathbf{a}_1) + \dots + \lambda^m f(\mathbf{a}_m) = \sum_{i=1}^{m} \sum_{j=1}^{n} A_{ij} \lambda^i b_j$$

and as $\{\mathbf{b}_1, \dots, \mathbf{b}_n\}$ is a basis of **W** we have

$$\mu^j = \sum_{i=1}^{m} A_{ij} \lambda^i \quad (j = 1, \dots, n)$$

or in matrix from

$$\begin{bmatrix} \mu^1 \\ \vdots \\ \mu^n \end{bmatrix} = A \begin{bmatrix} \lambda^1 \\ \vdots \\ \lambda^m \end{bmatrix}$$

Since $\boldsymbol{x} = x_k \boldsymbol{e_k}$; $T\boldsymbol{x} = x_k T \boldsymbol{e_k}$, the operator T is unambiguosly defined if its effect on each basis vector $\boldsymbol{e_k}$ of a basis $\boldsymbol{e}$ is known. Now $T\boldsymbol{e_k}$ is some definite vector of the space V and is thus expressible as a unique(except for order) linear combination of the basis vectors $\boldsymbol{e_k}$. Thus we have

$$T\boldsymbol{e_k} = a_{ik}\, \boldsymbol{e_i} \qquad \dots(1.17.2)$$

and with each linear operator T we can uniquely associate a square matrix,

$T \to A = (a_{ik})$ when once a basis $\boldsymbol{e}$ has been choosen.

Consider the operator equation

$$y = Tx \qquad ...(1.17.3)$$

On choosing a basis $\boldsymbol{e}$, we have

$$y_i\, \boldsymbol{e}_i = y = T\boldsymbol{x} = Tx_k\, \boldsymbol{e}_k = x_k T \boldsymbol{e}_k = x_k a_{ik}\, \boldsymbol{e}_i$$

and because of the linear independence of the $\boldsymbol{e}_i$ we obtain $y_i = a_{ik} x_k$ or in matrix notation $y = A\,x$, where x and y are the columns of the components x_i and y_i repectively of the vectors $\boldsymbol{x}$ and $\boldsymbol{y}$ relative to the basis $\boldsymbol{e}$. It is suggestive to write equation (1.17.3) also in matrix notation, taking $\boldsymbol{e}$ to be the column of basis vectors $\boldsymbol{e}_k$. We obtain (1.17.2)

$$T\boldsymbol{e} = A^T \boldsymbol{e} \qquad ...(1.17.4)$$

where A^T is the transpose of A. Note that x is transformed by A while e is transformed by A^T.

Example(1):

Consider the ordinary space R^3 of three-dimensional vectors. If $\boldsymbol{x}$ is an abitrary vector of R^3 and $\boldsymbol{A}$ is some definite vector, the vector product $y = \boldsymbol{A} \times \boldsymbol{x}$ may be regarded as the effect of an operator t_A dependent on $\boldsymbol{A}$, which maps the vector $\boldsymbol{x}$ into $y = t_A \cdot \boldsymbol{x} = \boldsymbol{A} \times \boldsymbol{x}$. To obtain the matrix representation of t_A relative to the basis $\boldsymbol{i}, \boldsymbol{j}, \boldsymbol{k}$, say, we make t_A act on each of them and write the image as a linear combination of $\boldsymbol{i}, \boldsymbol{j}, \boldsymbol{k}$ in accordance with equation(1.17.2).

Now

$$\begin{aligned} t_A \boldsymbol{i} &= \boldsymbol{A} \times \boldsymbol{i} = (A_1\, \boldsymbol{i} + A_2\, \boldsymbol{j} + A_3\, \boldsymbol{k}) \times \boldsymbol{i} = +A_3\, \boldsymbol{j} - A_2\, \boldsymbol{k} \\ t_A \boldsymbol{j} &= \boldsymbol{A} \times \boldsymbol{j} = (A_1\, \boldsymbol{i} + A_2\, \boldsymbol{j} + A_3\, \boldsymbol{k}) \times \boldsymbol{j} = -A_3\, \boldsymbol{i} + A_1\, \boldsymbol{k} \\ t_A \boldsymbol{k} &= \boldsymbol{A} \times \boldsymbol{k} = (A_1\, \boldsymbol{i} + A_2\, \boldsymbol{j} + A_3\, \boldsymbol{k}) \times \boldsymbol{k} = +A_2\, \boldsymbol{i} - A_1\, \boldsymbol{j} \qquad ...(1.17.5) \end{aligned}$$

so that we have

$$t_A = \begin{pmatrix} 0 & -A_3 & A_2 \\ A_3 & 0 & -A_1 \\ -A_2 & A_1 & 0 \end{pmatrix} = S_A(\text{say}) \qquad ...(1.17.6)$$

where S_A is a skew-symmetric matrix representing t_A obtained by transposing the matrix of transformation on the right of equation(1.17.5). It is easy to verify that in the corresponding matrix equation $y = S_A x$, the column vector y has the components

$(A_2x_3 - A_3x_2, A_3x_1 - A_1x_3, A_1x_2 - A_2x_1)$ as required.

Example (2):

Consider now the operation R of rotation of a vector $\boldsymbol{x}$ into the vector $\boldsymbol{y}$ about an axis passing through the origin of R^3. If $\boldsymbol{i}, \boldsymbol{j}, \boldsymbol{k}$ are the unit vectors along the axes of a rectangular cartesian frame, R rotates these vectors into $\boldsymbol{i}'; \boldsymbol{j}'; \boldsymbol{k}'$ which again form another rectangular cartesian frame. If $l_k = cos\,\alpha_k$, $m_k = cos\,\beta_k$, n_k = $\cos\gamma_k$, (k = 1,2,3) are the direction cosines respectively of $\boldsymbol{i}'; \boldsymbol{j}'$ and $\boldsymbol{k}'$, we have

$$\mathrm{R}\boldsymbol{i} = \boldsymbol{i}' = l_1\,\boldsymbol{i} + l_2\,\boldsymbol{j} + l_3\,\boldsymbol{k}$$

$$\mathrm{R}\boldsymbol{j} = \boldsymbol{j}' = m_1\,\boldsymbol{i} + m_2\,\boldsymbol{j} + m_3\,\boldsymbol{k}$$

$$\mathrm{R}\boldsymbol{k} = \boldsymbol{k}' = n_1\,\boldsymbol{i} + n_2\,\boldsymbol{j} + n_3\,\boldsymbol{k}$$

so that the matrix representation A of the operator R is given by

$$\mathrm{R} \to A = \begin{pmatrix} l_1 & m_1 & n_1 \\ l_2 & m_2 & n_2 \\ l_3 & m_3 & n_3 \end{pmatrix} \qquad ...(1.17.7)$$

which is called a rotation matrix. Some properties of A may be immediately noted. Since the elements of A are direction cosines, we have

(a) $l_1^2 + l_2^2 + l_3^2 = 1$ and similar relations for other columns

(b) $l_1m_1 + l_2m_2 + l_3m_3 = 0$ etc. for any pair of columns

(c) $l_1^2 + m_1^2 + n_1^2 = 1$ etc. for each row

(d) $l_1l_2 + m_1m_2 + n_1n_2 = 0$ etc. for any pair of rows.

It is easily seen that the relations (a) and (b) for columns amount to

$$A^T A = E \qquad ...(1.17.8)$$

while the relations (c) and (d) for the rows are equivalent to

$$AA^T = E \qquad ...(1.17.9)$$

Taking determinants of equation(1.17.8), we have

$det\, A^T \cdot det\, A = 1$ or $(det\, A)^2 = 1$ since $det\, A = det\, A^T$.

Thus $det\, A = \pm 1$ only so that A^{-1} exists and $A^T = A^{-1}$. Multiplying this on the left by A, we obtain equation(1.17.9). Similarly (1.17.9) yields equation(1.17.8).

A matrix A, satisfying equation (1.17.8) or equation (1.17.9) is called an orthogonal matrix and its determinant can only be +1 or –1. We will now show that the orthogonal matrix A of equation (1.17.7) must have determinant +1 only. The determinant of A is a continuous function of its elements which being direction cosines are themselves continuous functions of the angles of mutual orientation of the frames considered .Thus, $det\, A$ can change only continuously as these angles are continuously varied. Since $det\, A = \pm 1$ only, a discontinuous jump from +1 to -1 is impossible if $det\, A$ was +1, before effecting a continuous variation of the angles. Since the orientation of the frame $(\mathbf{i}'\,\mathbf{j}'\,\mathbf{k}')$ obtained by the rotation can be brought about from an initial orientation when the two frames are coincident, by a continuous variation of the angles and since $det\, A = +1$ in this initial position, $det\, A$ continues to remain +1.

Thus, a matrix A such as the one in equation(1.17.7) which represents a rotation must be an orthogonal matrix with determinant +1 and is called a proper orthogonal matrix.

We shall show later that the converse is also true, namely that any orthogonal matrix with determinant +1 does represent a rotaion.

We now consider some special cases of the A of equation (1.17.7). We note first of all, that a rotation about an axis along which a direction (positive or negative) is specified is said to be positive if a right-hand screw which is given the rotation in question advances along the positive direction of the axis.

1. Positive Rotation About the z-axis:

   Let R be a positive rotation through an angle θ about the positive direction of the z-axis. Then the required direction cosines are (see figure(1.1))

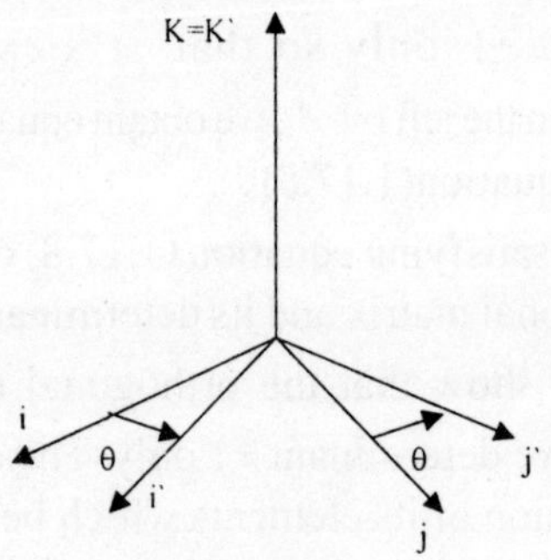

Figure (1.1)

$l_1 = cos\theta;\ l_2 = sin\theta;\ m_1 = -sin\theta;\ m_2 = cos\theta;\ n_3 = 1$ and the others are all zero. Thus,

$$R \to R(\boldsymbol{k},\theta) = \begin{pmatrix} cos\theta & -sin\theta & \cdot \\ sin\theta & cos\theta & \cdot \\ \cdot & \cdot & 1 \end{pmatrix} \qquad ...(1.17.10)$$

Similarly, the matrices for rotations about the other two axes are easily seen to be

$$R(\boldsymbol{j},\theta) = \begin{pmatrix} cos\theta & \cdot & sin\theta \\ \cdot & 1 & \cdot \\ -sin\theta & \cdot & cos\theta \end{pmatrix} \text{ and}$$

$$R(\boldsymbol{i},\theta) = \begin{pmatrix} 1 & \cdot & \cdot \\ \cdot & cos\theta & -sin\theta \\ \cdot & sin\theta & cos\theta \end{pmatrix}.$$

Here $R(a,\theta)$ represents the matrix for rotation about the unit vector a. The sign of θ must, of course, be reversed for a negative rotation.

(1.18) *Subspaces*

Definition: Let **V** be a vector space over K. A subset **W** of **V** is called a vector subspace, or simply a subspace of V if it satisfies the following conditions

If $\mathbf{a},\mathbf{b} \in \mathbf{W}$ and $\lambda \in K$, then $\mathbf{a}+\mathbf{b} \in \mathbf{W}$ and $\lambda\mathbf{a} \in \mathbf{W}$ or equivalently satisfies the condition

$\lambda\mathbf{a}+\mu\mathbf{b} \in \mathbf{W}$ for each $\mathbf{a},\mathbf{b} \in \mathbf{W}$ and $\lambda, \mu \in K$. Let **V** be a vector space over K. Let for some index set I, a set of vectors S= $\{v_i \in \mathbf{V} : i \in I\}$ is given , then the collection of all finite linear combinations of elements of S is a subspace of **V** called the span of S and is denoted by span (S). Given a vector space **V** over K and let $S \subseteq \mathbf{V}$. If span (S)=**V** and S is linearly independent then S is called a basis for **V.** Let $\mathbf{V}_1,\ldots, \mathbf{V}_n$ be n vector subspaces of **V**. If, for an arbitrary vector **a** of **V**, there are vectors $\mathbf{a}_i$ in $\mathbf{V}_i$ (i = 1, . . . ,n) such that

$$\mathbf{a} = \mathbf{a}_1 + \ldots + \mathbf{a}_n \qquad \text{...(1.18.1)}$$

then **V** is said to be the sum of $\mathbf{V}_1, \ldots, \mathbf{V}_n$ and we write

$$\mathbf{V} = \mathbf{V}_1 + \ldots + \mathbf{V}_n.$$

Furthermore, if an arbitrary element **a** of **V** can be expressed uniquely in the from (1.18.1), i.e., if $\mathbf{a} = \mathbf{a}_1 + \ldots + \mathbf{a}_n = \mathbf{a}'_1 + \ldots + \mathbf{a}'_n$ $(\mathbf{a}_i, \mathbf{a}'_i \in \mathbf{V}_i, i = 1, \ldots, n)$ implies $\mathbf{a}_1 = \mathbf{a}'_1$ (i = 1, . . . , n), then **V** is said to be the direct sum of $\mathbf{V}_1,\ldots,\mathbf{V}_n$ and we wirte

$$\mathbf{V} = \mathbf{V}_1 \oplus \ldots \oplus \mathbf{V}_n \text{ (direct sum).}$$

The necessary and sufficient condition that a vector space **V** be a direct sum of its subspaces $\mathbf{V}_1, \ldots, \mathbf{V}_n$ is that $\mathbf{V} = \mathbf{V}_1 + \ldots + \mathbf{V}_n$ and $\mathbf{V}_1 \cap \cdots \cap \mathbf{V}_n = \{0\}$

(1.19) ***Quotient space***

Let **V** be a vector space over K, and **W** a subspace of **V**. For **a**, **b** in **V**, we say that **a** and **b** are equivalent (with respect to **W**), if the condition **a-b** $\in$ **W** is satisfied, and we write **a** ~ **b**. The relation that **a** and **b** are equivalent satisfies the so-called equivalence laws:

1. Reflexive **a** ~ **a**
2. Symmetry If **a** ~ **b**, then **b** ~ **a**
3. Transitive If **a** ~ **b**, **b** ~ **c**, then **a** ~ **c**.

In general if a relation between the elements of a set X is defined, satisfying the equivalence laws above, then we call this relation an equivalence relation in X. When an equivalence relation ~ is defined in X, a subset C of X is called an equivalence class with respect to this equivalence relation if it has the following property: if **a**, **b** $\in$ C then **a** and **b** are equivalent and if **a** $\in$ C and **b** $\notin$ C then **a** and **b** are not equivalent. Let **a**$\in$X and set $C_a = \{\mathbf{b} \in X : \mathbf{a}\, C_a\, \mathbf{b}\}$. Then C_a is an equivalence class containing **a**. Conversely, let C be an arbitrary equivalence class, and let **a** $\in$ C. Then C = C_a. If C and C′ are two equivalence classes, then C = C′ or C $\cap$ C′ = $\varnothing$ holds. The set of equivalence classes is denoted by X /~, and is called the quotient set of X with respect to the equivalence relation.

Returning to the vector space **V**, where an equivalence relation **a** ~ **b** is defined with respect to a subspace **W**, the quotient set **V**/~ is denoted by **V/W**. For an arbitrary element **a** of **V**, we write the equivalence class containing **a** as $(\bar{\mathbf{a}})$. Then **a** ~ **b** is the same as $(\bar{\mathbf{a}}) = (\bar{\mathbf{b}})$, and we have

$$(\bar{\mathbf{a}}) = \{\mathbf{a} + \mathbf{w} : \mathbf{w} \in \mathbf{W}\}.$$

Now if **a** ~ **a**′, **b** ~ **b**′, and $\lambda \in$ K, we can easily see that **a** + **b** ~ **a**′+ **b**′ and λ**a** ~ λ**a**′ hold. Hence we can define the sum of two elements of **V/W** and the product of an element of **V/W** by an element of K by

$$(\bar{\mathbf{a}}) + (\bar{\mathbf{b}}) = (\overline{\mathbf{a}+\mathbf{b}})$$
$$\lambda(\bar{\mathbf{a}}) = (\overline{\lambda \mathbf{a}}).$$

In fact, the definitions given above do not depend on the choice of representatives **a,b** for the equivalence classes $(\bar{\mathbf{a}})$, $(\bar{\mathbf{b}})$. If we define addition and scalar multiplication as above, then **V/W** becomes a vector space over K. The vector space **V/W** is called the quotient vector space of the vector space **V** by its subspace **W**.

(1.20) *Dual spaces*

Let **V** be a vector space over K. A map f from **V** into K satisfying

$$f(\mathbf{a} + \mathbf{b}) = f(\mathbf{a}) + f(\mathbf{b}) \ (\mathbf{a}, \mathbf{b} \in \mathbf{V}), \text{ and}$$
$$f(\lambda \mathbf{a}) = \lambda f(\mathbf{a}) \ (\lambda \in K, \mathbf{a} \in \mathbf{V})$$

is called a linear form or a linear function on **V**. A linear function is nothing but a linear map from **V** into the 1-dimensional vector space K . Let $\mathbf{e}_1,..., \mathbf{e}_n$ be a basis in an n-dimensional space **V**, since every vector **a** can be represented in the form $\mathbf{a} = \lambda^1 \mathbf{e}_1 +...+ \lambda^n \mathbf{e}_n$. The linear function f on **V** imply that

$f(\mathbf{a}) = f(\lambda^1 \mathbf{e}_1 +...+ \lambda^n \mathbf{e}_n) = \lambda^1 f(\mathbf{e}_1)+...+ \lambda^n f(\mathbf{e}_n) = \lambda^1 \mathbf{a}_1 +... + \lambda^n \mathbf{a}_n$, where the coefficient $\mathbf{a}_1,...,\mathbf{a}_n$ which determine the linear function f are given by $\mathbf{a}_1 = f(\mathbf{e}_1),..., \mathbf{a}_n = f(\mathbf{e}_n)$ It is clear from above that given a basis $\mathbf{e}_1,..., \mathbf{e}_n$ every n-tuple $\mathbf{a}_1,...,\mathbf{a}_n$ determines a unique linear function. Let the set of all linear functions on **V** be denoted by **V***. For f, g $\in$ **V***, $\lambda \in$ K, we define maps $f+ g$, λf from **V** into K by

$$(f+ g)(\mathbf{a}) = f(\mathbf{a}) + g(\mathbf{a}), \qquad (\lambda f)(\mathbf{a}) = \lambda(f(\mathbf{a})) \ (\lambda \in K, \mathbf{a} \in \mathbf{V}).$$

It is easy to see that $f+ g$ and λf are linear functions on **V**. Hence addition between elements of **V*** and scalar multiplication with elements of K and elements of **V*** are defined, and with this addition and scalar multiplication **V*** becomes a vector space over K. The vector space **V*** is called the dual space of **V**. The linear function which associates to each element $\mathbf{0} \in \mathbf{V}$ the scalar $0 \in K$ is called **0**-vector in **V***. We shall denote the value of a linear function f at a point $\mathbf{a} \in \mathbf{V}$ by $(f, \mathbf{a})$.

Now let **V** have dimension n and let $\{\mathbf{e}_1, \ldots, \mathbf{e}_n\}$ be a basis of **V**. For an arbitrary element $\mathbf{a} = \lambda^1 \mathbf{e}_1 + \ldots + \lambda^n \mathbf{e}_n$ of **V**, we set the ith component λ^i of **a** equal to f^i (**a**). It is easy to see that the map $\mathbf{a} \to f^i(\mathbf{a})$ defines a linear function on **V** for each i = 1, . . . , n, and satisfies

$$(f^i, \mathbf{e}_j) = \delta^i_j = \begin{cases} 1 & \text{when } i = j \\ 0 & \text{when } i \neq j \end{cases} \quad (i, j = 1, \ldots, n) \qquad (1.20.1)$$

If $\mathbf{e}_1, \ldots, \mathbf{e}_n$ is a basis in **V** , then $(f, \mathbf{e}_k) = f(\mathbf{e}_k)$ gives the number $\mathbf{a}_k$ which determine the linear function $f \in \mathbf{V}^*$.

The set of linear functions $f^1, \ldots, f^n$ is a linearly independent vectors of **V***. In fact, if $\alpha_1 f^1 + \ldots + \alpha_n f^n = \mathbf{0}$ $(\alpha_i \in K)$, then for an arbitrary $\mathbf{a} \in \mathbf{V}$ we have

$$\mathbf{0} = \mathbf{0}(\mathbf{a}) = (\alpha_1 f^1 + \ldots + \alpha_n f^n)(\mathbf{a}) = \sum_{i=1}^{n} \alpha_i f^i(\mathbf{a}) .$$

Hence, in particular for $\mathbf{a} = (\mathbf{e}_1, \ldots, \mathbf{e}_n)$, we have $\sum_{i=1}^{n} \alpha_i f^i(\mathbf{e}_i) = \mathbf{0}$. Thus we get $\alpha_i = 0$ $(i = 1, \ldots, n)$, and so $f^1, \ldots, f^n$ are linearly independent. Now let f be an arbitrary element of **V***, and set $f(\mathbf{a}_i) = \alpha_i$. We shall show

$f = \sum_{i=1}^{n} \alpha_i f^i$. Set $\sum_{i=1}^{n} \alpha_i f^i = g$. Then we have

$g(\mathbf{e}_i) = \alpha_i = f(\mathbf{e}_i)$. Now for $\mathbf{a} \in V$, let

$\mathbf{a} = \sum_{i=1}^{n} \lambda^i e_i$. Then $g(\mathbf{a}) = \sum_{i=1}^{n} \lambda^i g(\mathbf{e}_i) = \sum_{i=1}^{n} \lambda^i f(\mathbf{e}_i) = f(\mathbf{a})$.

Hence $f = g$. We have just proved that $\{f^1, \ldots, f^n\}$ is a basis of

$\mathbf{V}^*$. Hence the dual space $\mathbf{V}^*$ of the n-dimensional vector space $\mathbf{V}$ is also n dimensional. When condition (**1.20.1**) is satisfied for a basis $\{\mathbf{e}_1,...,\mathbf{e}_n\}$ of $\mathbf{V}$ and a basis $\{f^1, \ldots, f^n\}$ of $\mathbf{V}^*$, $\{f^1, \ldots, f^n\}$ is called a dual basis of $\{\mathbf{e}_1,...,\mathbf{e}_n\}$. For a given basis $\{\mathbf{e}_1,...,\mathbf{e}_n\}$ of $\mathbf{V}$ there is one and only one dual basis $\{f^1, \ldots, f^n\}$. Moreover for an arbitrary $\mathbf{a}\in\mathbf{V}$ we have $\mathbf{a}=\lambda^i\mathbf{e}_i$ then $(f^j,\mathbf{a}) = (f^j,\lambda^i\mathbf{e}_i) = \lambda^i(f^j,\mathbf{e}_i) = \lambda^i\delta_i^j=\lambda^j$

Hence, the coordinates λ^j of a vector $\mathbf{a}\in\mathbf{V}$ in the basis $\mathbf{e}_1,...,\mathbf{e}_n$ can be computed from the formula $\lambda^j=(f^j,\mathbf{a})$ where f^j is the basis dual tothe basis $\mathbf{e}_i$ .Similarly, if $f\in\mathbf{V}^*$ and $f=\alpha_i f^i$ then $\alpha_i=(f,\mathbf{e}_i)$. Now let $\mathbf{e}_1,...,\mathbf{e}_n$ and $f^1, \ldots, f^n$ be dual basis , we shall express the number $(f, \mathbf{a})$ in terms of the coordinates of the vectors f and $\mathbf{a}$ with respect to the basis $\mathbf{e}_1,...,\mathbf{e}_n$ and $f^1, \ldots, f^n$ respectively. Thus, let $\mathbf{a} = \lambda^1\mathbf{e}_1 + \ldots + \lambda^n\mathbf{e}_n$ and $f = \alpha_1 f^1 + \ldots + \alpha_n f^n$, then $(f,\mathbf{a}) = (\alpha_i f^i, \lambda^j \mathbf{e}_j) = \alpha_i \lambda^j (f^i, \mathbf{e}_j) = \alpha_i \lambda^j \delta_j^i = \alpha_i \lambda^i$. Thus if $\mathbf{e}_1,...,\mathbf{e}_n$ is a basis in $\mathbf{V}$ and $f^1, \ldots, f^n$ its dual basis in $\mathbf{V}^*$ then $(f,\mathbf{a}) = \alpha_1 \lambda^n + \ldots + \alpha_n \lambda^n$ where $\lambda^1 \ldots \lambda^n$ are the coordinates of $\mathbf{a}\in\mathbf{V}$ relative to the basis $\mathbf{e}_1,...,\mathbf{e}_n$ and $\alpha_1, \ldots, \alpha_n$ are the coordinates of $f\in\mathbf{V}^*$ relative to the basis $f^1, \ldots, f^n$. Note that for an arbitrary basis $\mathbf{e}_1,...,\mathbf{e}_n$ and $f^1, \ldots, f^n$ in $\mathbf{V}$ and $\mathbf{V}^*$ respectively we have $(f,\mathbf{a}) = A_{ij}\alpha_i\lambda^j$ where $A_{ij} = (f^i, \mathbf{e}_j)$. Now as $\mathbf{V}^*$ is also a vector space, we can consider the dual space $(\mathbf{V}^*)^*$ of $\mathbf{V}^*$. For $\mathbf{a}\in\mathbf{V}$ set $\mathbf{a}'(f) = f(\mathbf{a})$ $(f\in\mathbf{V}^*)$. It is easy to see that $\mathbf{a}'$ is a linear function on $\mathbf{V}^*$. Moreover $\mathbf{a}\mapsto\mathbf{a}'$ is a linear map from $\mathbf{V}$ to $(\mathbf{V}^*)^*$. Let $\{\mathbf{e}_1, \ldots, \mathbf{e}_n\}$, be a basis of $\mathbf{V}$ and $\{f^1, \ldots, f^n\}$ the basis dual to $\{\mathbf{e}_1, \ldots, \mathbf{e}_n\}$. As $\mathbf{e}'_i(f^j) = \delta^j_i$, $\{\mathbf{e}'_1,...,\mathbf{e}'_n\}$ is the basis dual to $\{f^1, \ldots f^n\}$. Hence, in particular, $\{\mathbf{e}'_1, \ldots, \mathbf{e}'_n\}$ is a basis of $(\mathbf{V}^*)^*$, and an arbitrary element $\mathbf{a}'$ of $(\mathbf{V}^*)^*$ is of the form $\mathbf{a}' = \sum_{i=1}^{n}\lambda^i\mathbf{e}'_i$. So, if we identify $\mathbf{a} = \sum_{i=1}^{n} \lambda^i \mathbf{e}_i$ with $\mathbf{a}'$, then we can consider the dual space $(\mathbf{V}^*)^*$ of $\mathbf{V}^*$ to be $\mathbf{V}$ itself.

Now let f be a linear map from a vector space $\mathbf{V}$ over K to a vector space $\mathbf{W}$ over K. For $g \in \mathbf{W}^*$, we let $(t_f g)(\mathbf{a}) = g(f(\mathbf{a}))$, $\mathbf{a} \in \mathbf{V}$, then $t_f g$ is a linear function on $\mathbf{V}$ i.e., $t_f g \in \mathbf{V}^*$. Hence $g \mapsto t_f g$, is map t_f from $\mathbf{W}^*$ to $\mathbf{V}^*$. It is easy to check that t_f is a linear map. The linear map t_f is called the dual map of the linear map f from $\mathbf{V}$ to $\mathbf{W}$.

(1.21) *Bilinear form*

(A) Let $\mathbf{V}$and $\mathbf{W}$ be vector spaces over K. A map $\eta : \mathbf{V} \times \mathbf{W} \to K$ which associates to each element $(\mathbf{a}, \mathbf{a}') \in \mathbf{V} \times \mathbf{W}$ an element $\eta(\mathbf{a}, \mathbf{a}')$ is called bilinear form on $\mathbf{V} \times \mathbf{W}$ if the following conditions are satisfied

(1) $\eta(\mathbf{a}+\mathbf{b}, \mathbf{a}') = \eta(\mathbf{a}, \mathbf{a}') + \eta(\mathbf{b}, \mathbf{a}')$,

(2) $\eta(\mathbf{a}, \mathbf{a}'+\mathbf{b}') = \eta(\mathbf{a}, \mathbf{a}') + \eta(\mathbf{a}, \mathbf{b}')$

(3) $\eta(\lambda\mathbf{a}, \mathbf{a}') = \eta(\mathbf{a}, \lambda\mathbf{a}') = \lambda\eta(\mathbf{a}, \mathbf{a}')$

$(\mathbf{a}, \mathbf{b} \in \mathbf{V}, \mathbf{a}', \mathbf{b}' \in \mathbf{W}\ ; \lambda \in K)$.

In particular if $\mathbf{V}=\mathbf{W}$, η is called a bilinear form on $\mathbf{V}$.

Let η be a bilinear form on $\mathbf{V} \times \mathbf{W}$. If $\eta(\mathbf{x}, \mathbf{a}') = 0$ for all $\mathbf{x} \in \mathbf{V}$ implies $\mathbf{a}' = 0$, and if $\eta(\mathbf{a}, \mathbf{y}) = 0$ for all $\mathbf{y} \in \mathbf{W}$ imples $\mathbf{a} = 0$, then η is called nondegenerate bilinear form on $\mathbf{V} \times \mathbf{W}$.

For $\mathbf{a} \in \mathbf{V}$ and $f \in \mathbf{V}^*$, let $< \mathbf{a}, f > = f(\mathbf{a})$. then the map $(\mathbf{a}, f) \mapsto < \mathbf{a}, f >$ is a nondegenerate bilinear form on $\mathbf{V} \times \mathbf{V}^*$.

Now let η be a bilinear form on $\mathbf{V}$, i.e., on $\mathbf{V} \times \mathbf{V}$. If for any $\mathbf{a}, \mathbf{b} \in \mathbf{V}$, $\eta(\mathbf{a}, \mathbf{b}) = \eta(\mathbf{b}, \mathbf{a})$,

holds, then η is called a symmetric bilinear form on $\mathbf{V}$, and if $\eta(\mathbf{a}, \mathbf{b}) = -\eta(\mathbf{b}, \mathbf{a})$ holds,

then η is called an alternating or skew-symmetric bilinear form on $\mathbf{V}$.

Let the dimension of $\mathbf{V}$ be n and let $\{\mathbf{a}_1, ..., \mathbf{a}_n\}$ be a basis of $\mathbf{V}$.

Let η be a bilinear form on **V** and set

$\eta(\mathbf{a}_i, \mathbf{a}_j) = n_{ij}$ $(i, j = 1 ..., n)$

Then we have

$$\eta(\mathbf{a}, \mathbf{b}) = \sum_{i,j=1}^{n} \eta_{ij}\lambda^i\mu^j \quad \text{for } \mathbf{a} = \sum_{i=1}^{n} \lambda^i \mathbf{a}_i, \ \mathbf{b} = \sum_{i=1}^{n} \mu^i \mathbf{a}_i$$

It is easy to prove that for a bilinear form η on **V**

(1) η is symmetric if and only if $\eta_{ij} = \eta_{ji}$ (i, j = 1, ... n)

(2) η is alternating if and only if $\eta_{ji} = -\eta_{ij}$ (i, j = 1, ... n)

It is easy to show that a bilinear form α on **V** is nondegenerate if and only if the determinant det (α_{ij}) corresponding to α is not 0. Consider now the case where K = R, i.e., where **V** is a real vector space. A nondegenerate symmetric bilinear form on **V** is called an inner product on **V**. That is, a bilinear form η on a real vector space **V** is an inner product if the two conditions (1) $\eta(\mathbf{x}, \mathbf{y}) = \eta(\mathbf{y}, \mathbf{x})$ and (2) $\eta(\mathbf{x}, a) = 0$ for all $\mathbf{a} \in \mathbf{V}$ implies $\mathbf{x} = 0$, hold. If a symmetric bilinear form η on **V** satisfies, $\eta(\mathbf{x}, \mathbf{x}) > 0$ for $\mathbf{x} \in$ V, $\mathbf{x} \neq 0$, then η is called a positive inner product on **V**. Now consider the case where K = C, i.e., where **V** is a complex vector space. If a map from **V** × **V** to C satisfies the conditions:

1. h $(\mathbf{x} + \mathbf{y}, \mathbf{z})$ = h $(\mathbf{x}, \mathbf{z})$ + h $(\mathbf{y}, \mathbf{z})$
2. h $(\mathbf{x}, \mathbf{y}) = \overline{h(y, x)}$ $(\mathbf{x}, \mathbf{y}, \mathbf{z} \in \mathbf{V}; \lambda \in C)$
3. $h(\lambda\mathbf{x}, \mathbf{y}) = \lambda h(\mathbf{x}, \mathbf{y})$

then h is called a hermitian form on **V**.

If h $(\mathbf{x}, \mathbf{a}) = 0$ for all $\mathbf{a} \in \mathbf{V}$ implies $\mathbf{x} = 0$ then h is called a hermitian inner product on the complex vector space **V**. If h is a hermitian form, then h $(\mathbf{x}, \mathbf{x})$ is a real number. If h satisfies h $(\mathbf{x}, \mathbf{x}) > 0$ for $\mathbf{x} \in$ V, $\mathbf{x} \neq 0$, then h is called a positive hermitian inner product and, we called V as Unitary Space.

1(e) Differential Calculus of functions of several variables

(1.22) Let R^n donote the set of all n-tuples of real numbers $(\mathbf{x}^1, \ldots, \mathbf{x}^n)$. For a point $\mathbf{x} = (\mathbf{x}^1,\ldots,\mathbf{x}^n)$ of $\mathbf{R}^n$, let $|\,\mathbf{x}\,| = \underset{i=1\ldots n}{\text{Max}} |\mathbf{x}^i|$, For two points $\mathbf{x}$, $\mathbf{y}$ in R^n, let $\mathbf{x} - \mathbf{y} = (\mathbf{x}^1 - \mathbf{y}^1, \ldots, \mathbf{x}^n - \mathbf{y}^n)$, and let $d(\mathbf{x}, \mathbf{y}) = |\mathbf{x} - \mathbf{y}|$, then d gives a metric on $\mathbf{R}^n$. Hence a topology is defined in $\mathbf{R}^n$ and $\mathbf{R}^n$ becomes a topological space. This topological space is called the n-dimensional real space, and is also denoted by $\mathbf{R}^n$, $\mathbf{R}^n$ is a complete metric space. For $\mathbf{x} \in \mathbf{R}^n$ and $r > 0$, let

$Q(\mathbf{x}\,;\,r) = \{\mathbf{y} \in \mathbf{R}^n \mid |\,\mathbf{x} - \mathbf{y}| < r\}$, $Q(\mathbf{x}\,;\,r)$ is called a *cube* of half - width r centered at $\mathbf{x}$. A subset $\mathbf{O}$ of $\mathbf{R}^n$ is an open set of $\mathbf{R}^n$ if and only if for an arbitrary point $\mathbf{x}$ of $\mathbf{O}$, there is an $r > 0$ such that $Q(\mathbf{x}\,;\,r) \subset \mathbf{O}$. For a point $\mathbf{x} = (\mathbf{x}^1,\ldots,\mathbf{x}^n)$ in $\mathbf{R}^n$, let $p^i(\mathbf{x}) = \mathbf{x}^i$ $(i = 1,\ldots n)$.

$p^1,\ldots, p^n$ are continuous function from $\mathbf{R}^n$ to $\mathbf{R} = \mathbf{R}^1$. We call this n-tuple of functions $(p^1,\ldots, p^n)$ the standard coordinate system of $\mathbf{R}^n$. The notation $(p^1,\ldots, p^n)$ denotes the standard coordinate system in R^n, but it is conventional and often convenient to use the same notation $\mathbf{x} = (\mathbf{x}^1,\ldots, \mathbf{x}^n)$ to denote a "general point" of $\mathbf{R}^n$.

A map f from a subset A of $\mathbf{R}^n$ to $\mathbf{R}$ is called a function $f(\mathbf{x}^1,\ldots, \mathbf{x}^n)$ on n-variables $\mathbf{x}^1,\ldots, \mathbf{x}^n$ defined on A (or with domain A). $\mathbf{R}^n$ can also be considered as n-dimensional vector space over $\mathbf{R}$. That is, for points $\mathbf{x} = (\mathbf{x}^1,\ldots,\mathbf{x}^n)$, $\mathbf{y} = (\mathbf{y}^1,\ldots, \mathbf{y}^n)$ of $\mathbf{R}^n$ and for $\lambda \in \mathbf{R}$, we define the sum of elements in $\mathbf{R}^n$ and the scalar multiplication by $\mathbf{x}+\mathbf{y} = (\mathbf{x}^1 + \mathbf{y}^1,\ldots,\mathbf{x}^n + \mathbf{y}^n)$, $\lambda\mathbf{x} = (\lambda\mathbf{x}^1,\ldots,\lambda\mathbf{x}^n)$, and thus $\mathbf{R}^n$ becomes an n-dimensional vector space over $\mathbf{R}$. Let f be a function defined on an open set U of $\mathbf{R}^n$ and let $\mathbf{x} \in U$. Let $\mathbf{y}$ be a point in a neighborhood of $\mathbf{0} = (\mathbf{0},\ldots,\mathbf{0})$ If $f(\mathbf{x}+\mathbf{y}) - f(\mathbf{x})$ can be written as $f(\mathbf{x} + \mathbf{y}) - f(\mathbf{x}) = \sum_{i=1}^{n} A_i \mathbf{y}^i + c(\mathbf{y})$ [here A_i is a constant independent of $\mathbf{y}$, $c(\mathbf{y})$ is a function of $\mathbf{y}$] and if $c(\mathbf{y})/|\,\mathbf{y}|$ approaches 0 as $\mathbf{y}$ approaches the origin 0, then we say that f is totally differentiable at **x**. If f is totally differentiable at **x** then f is differentiable with respect to each variable $\mathbf{x}^i$ at **x** and

$\frac{\partial f}{\partial \mathbf{x}^i}(\mathbf{x}^1, ..., \mathbf{x}^n) = A_i$ (i = 1, ..., n) hold. If f is totally differentiable at all points of U. then we say that f is totally differentiable on U. Let f be a function defined on an open set U of $\mathbf{R}^n$. If f is differentiable at all points of U with respect to each variable $\mathbf{x}^i$, and if each partial derivative $\frac{\partial f}{\partial \mathbf{x}^i}$ (i=1,...,*n*) is continuous on U, then f is called a continuously differentiable function or a C^1-function on U. If for a positive integer r, the rth order partial derivatives

$\frac{\partial^{\alpha_1+...+\alpha_n} f}{(\partial \mathbf{x}^1)^{\alpha_1}...(\partial \mathbf{x}^n)^{\alpha_n}}$ $(\alpha_i \geq 0,\ \alpha_1 + ... + \alpha_n \leq r)$ all exist and are continuous on U, then f is called an r-times continuously differentiable function or a C^r-function on U, or we say that f is (of class) C^r on U. If f is of class C^r on U for every positive integer r then we say that f is C^∞ function on U or that f is (of class) C^∞ on U. We also say that f is C^0 function on U or that f is (of class) C^0 if f is continuous on U.

Smooth Functions

Definition

Let $U \subset R^n$ be a set, and let $F: U \rightarrow R^m$ be a map. We say F is smooth if

(1) The set U is open in R^n;
(2) All partial derivatives of F of all orders exist and are continuous.

We can write F using coordinate functions as

$$F(x) = \begin{pmatrix} F_1(x) \\ \vdots \\ F_m(x) \end{pmatrix},$$

where $x = \begin{pmatrix} x_1 \\ \vdots \\ x_m \end{pmatrix}$ and $F_1, \ldots, F_m : U \to R$ are smooth functions.

The **Jacobian Matrix** of F is the matrix of partial

derivatives $$DF = \begin{pmatrix} \frac{\partial F_1}{\partial x_1} & \cdots & \frac{\partial F_1}{\partial x_n} \\ \vdots & \vdots & \vdots \\ \frac{\partial F_m}{\partial x_1} & \cdots & \frac{\partial F_m}{\partial x_n} \end{pmatrix}$$

The openness of the set U in the above definition will often be unstated but will be assumed nonetheless. The following definition is the smooth analog of the notion of homeomorphism.

(1.23) ***Definition (Derivative in Banach space)***

Let E and F denote the Banach space and Let U be an open subset of E. Then the function f: U $\to$ Fis differentiable at p $\in$ U if there exists a continuous linear map T $\in$L(E, F) so that for all x $\in$E the limit

$$\lim_{X \to 0} \frac{\|f(p+x) - f(p) - T(x)\|}{\|x\|} = 0$$

where $\|\cdot\|$ denotes the norm in E or F.

Thus for x sufficiently small in E. We have the approximation for a differentiable function

$$f(p + x) = f(p) + T(x) + \| x \| \varepsilon(x)$$

where ε(x) is a function such that $\lim_{X \to 0} \varepsilon(x) = 0$.

If such a continuous linear map T exists , then T is called the derivative of f at p $\in$U and is denoted by Df(p), f′ (p) or df(p).

Thus

$$Df(p)x = T(x) \quad \text{for all } x \in E.$$

(1.24) ***Theorem***

Let f: U → F be differentiable at $p \in U$, L(E, F) be the Banach space of continuous linear maps and let $T_1, T_2 \in L(E, F)$, satisfy

$$\lim_{X \to 0} \frac{\|f(p+x) - f(p) - T_i(x)\|}{\|x\|} = 0$$

for i = 1,2. and for all $x \in E$. Then $T_1 = T_2 = Df(p)$.

Further for all $x \in E$

$$Df(p).x = \lim_{t \to 0} \frac{1}{t}[f(p+tx) - f(p)]$$

Proof: For $x \neq 0$, $x \in E$, let d(x) = f(p + x) - f(p). Then we have

$$\frac{\|T_1(x) - T_2(x)\|}{\|x\|} = \lim_{t \to 0} \frac{\|T_1(tx) - T_2(tx)\|}{\|tx\|}$$

$$= \lim_{t \to 0} \frac{\|T_1(tx) - d(tx) + d(tx) - T_2(tx)\|}{\|tx\|}$$

$$= \lim_{t \to 0} \|\varepsilon_1(tx) + \varepsilon_2(tx)\|$$

$$= 0$$

$$\Rightarrow T_1(x) = T_2(x) \ \forall x \in E$$

we also have $Df(p).\, x = \lim_{t \to 0} \frac{1}{t} Df(p)(tx) + \lim_{X \to 0} \frac{1}{t} \|tx\| \in (tx)$

$$= \lim_{t \to 0} \frac{1}{t} [f(p+tx) - f(p)]$$

(1.25) ***Example***

(1) Let f : E → F be a continuous linear map. Then Df(p) = f

for all $p \in E$, for let $x \in E$, then $f(p + x) - f(p) = f(x)$. Thus for $T = f$ we see

$$f(p + x) - f(p) - T(x) = 0.$$

So that the limit in the definition is 0 and by uniqueness $Df(p) = f$.

(2) Let $E = F = R$ and let $f : U \to F$ be differentiable at p in the usual calculus sense with

$$f'(p) = \lim_{t \to 0} \frac{1}{t} [f(p + t) - f(p)].$$

Then the continuous linear map $T : R \to R$ which takes $x \mapsto f'(t)x$ given by multiplication satisfies the above definition. Thus Df(p) has the | x | matrix (f′(p)) relative to the basis of R consisting of the number 1.

For a differential function $f : U \subset R \to R$ the usual interpretation of the derivative at $u_0 \in U$ is the slope of the line tangent to the graph of f at u_0. To generalize this, we interpret $Df(u_0) = f'(u_0)$ as a linear map acting on the vector $(u - u_0)$. Then we can say that $Df(u_0)$ is the unique linear map from R into R such that the mapping $g : U \to R$ given by

$$u \mapsto g(u) = f(u_0) + Df(u_0).(u - u_0)$$

is tangent to f at u_0. This version can be extended to vector valued mapping of a vector variable as follows.

(1.26) *Definition*

Let E, F be Banach space with maps f,g: $U \subset E \to F$ where U is an open set in E. We say f and g are tangent at $u_0 \in U$, if

$$\lim_{u \to u_0} \frac{\|f(u) - g(u)\|}{\|u - u_0\|} = 0$$

where $\| . \|$ represent norm on the appropriate space.

Now suppose that f is differentiable in U. Then the map

$$U \ni u \rightarrow Df(u) \in Clm(E, F)$$

is called the derivative of f, and will be denoted by Df.

Important Results of Calculus on Banach Space

(1.27) Let f be a function on a subset U of A into a Banach space E over A, and let u be an interior point of U such that f is differentiable at u. The derivative Df(u) of f at u is a member of Clm(A, E), which is identified with E itself through the isomorphism

$$Clm(A, E) \ni g \mapsto g(1) \in E$$

In particular, Df(u) is identified with Df(u).1, which is the derivative of f at u along the vector 1 in A, considered as a Banach space over itself. It is customary to write (df /du) for Df(u).1.

(1.28) A constant function f on a Banach space E over A into a Banach space F over A is evidently differentiable at each point x of E and Df(x) is the zero-member of Clm(E, F). A continuous linear map g of E into F is differentiable at each point x of E and Dg(x) = g.

(1.29) (Chain Rule) Let f be a function on an open subset U of a Banach space E over A into an open subset V of a Banach space F over A and let g be a function on V into a Banach space H over A such that f is differentiable at a point u of U and g is differentiable at f(u). Then $g \circ f$ is differentiable at u and

$$D(g \circ f)(u) = (Dg(f(u))) \circ Df(u).$$

(1.30) Let f and g be function on an open subset U of a Banach space E over A into a Banach space F over A such that f and g are differentiable in U. Let v be any vector in E and let a be any scalar in A. Then the functions

$$af: x \mapsto af(x),\ f + g : x \mapsto f(x) + g(x),\ h : x \mapsto x + v$$

on U are differentiable in U and for each u in U

$$D(af)(u) = a\,Df(u),$$

$$D(f + g)(u) = Df(u) + Dg(u)$$

and $$Dh(u) = id_E .$$

(1.31) Let n be a positive integer and for each i in $\underline{n}$ let F_i be a

Banach space over A. Let f be a function on an open subset U of a Banach space E over A into the product space $\prod_{i \in \underline{n}} F_i$ over A. For each j in $\underline{n}$ let pr_j be the projection of $\prod_{i \in \underline{n}} F_i$ onto F_j, let q_j be the canonical monomorphism of F_j into $\prod_{i \in \underline{n}} F_i$, and let $f_j = pr_j \circ f$. Then f is differentiable at a point u of U iff for each j in $\underline{n}$, f_j is differentiable at u. Moreover for each u in U and each x in X

$$Df(u).x = \sum_{j \in \underline{n}} q_j\left(Df_j(u).x\right)$$

$$=(Df_j(u).x \,;\, j \in \underline{n}).$$

(1.32) For each i in $\underline{n}$ let E_i is a Banach space over A. Put $E = E_1 \times ... \times E_n$, and let f be a function on a subset U of $E_1 \times ... \times E_n$ into a Banach space F over A. Let $u = (u_1,...,u_n)$ be an Interior Point of U, such that for each i in $\underline{n}$ the function

$$x_i \mapsto f(u_1,..., u_{i-1}, x_i, u_{i+1},...,u_n)$$

on E_i into F is differentiable at u_i; then the derivative $Df_i(u_i)$ of f_i at u_i is called the i-th partial derivative of f at u, denoted by $D_i f(u)$, and is a member of $Clm(E_i, F)$.

Let E be equal to A^n and let $(e_1,...,e_i,...,e_n)$ be the canonical basis of A^n. Then for each i in $\underline{n}$ $D_i f(u)$ is a member of Clm(A, F), and so $D_i f(u)$ can be identified with $D_i f(u).1$, which is identical with $D_{e_i} f(u)$.

(1.33) Let n be a positive integer and for each i in n let E_i be a Banach space over A. Let U be a subset of the product space

$$\prod_{i \in \underline{n}} E_i,$$

Let $u = (u_1,...,u_i,...,u_n)$ be an interior point of U and let f be a function on U into a Banach space F over A such that f is differentiable at u. Then for each j in n the partial derivative $D_j f(u)$ of f at u exists and for all

$$x = (x_1,\ldots,x_i,\ldots,x_n) \text{ in } \prod_{i\in\underline{n}} E_i$$

$$Df(u).x = \sum_{j\in\underline{n}} D_j f(u).x_j$$

(1.34) For each i in $\underline{n}$ let f_i be a function on a subset U of a Banach space E over A into a Banach space F_i over A, let u be an interior point of U such that f_i is differentiable at u. Let F be a subset of the product space $F_1 \times \ldots \times F_n$ such that $(f_1(u),\ldots,f_n(u))$ is an interior point of F and for each x in U, $(f_1(x),\ldots,f_n(x))$ is in F. Let g be a function on F into a Banach space H over A such that g is differentiable at $(f_1(u),\ldots,f_n(u))$. Then by (1.29), (1.31), and (1.33) the function h

$$U \ni x \mapsto g(f_1(x),\ldots,f_n(x)) \in H$$

is differentiable at u and for each x in U

$$Dh(u).x = \sum_{i\in\underline{n}} D_i\, g(f_1(u),\ldots,f_n(u))(Df_i(u).x).$$

(1.35) For each i in n let f_i be a function on a subset U_i of a Banach space E_i over A into a Banach space F_i over A such that f_i is differentiable at an interior point u_i of U_i. Let f be the function

$$\prod_{i\in\underline{n}} U_i \ni (x_i; i\in\underline{n}) \mapsto (f_i(x_i); i\in\underline{n}) \in \prod_{i\in\underline{n}} F_i\ .$$

Then by (1.11) and (1.9) f is differentiable at $(u_i;\ i\in\underline{n})$ and for each $(x_i;\ i\in\underline{n})$ in $\prod_{i\in\underline{n}} E_i$

$$Df(u_1,\ldots,u_n).(x_1,\ldots,x_n) = (Df_1(u_1).x_1,\ldots,\ Df_n(u_n).x_n)$$

(1.36) Let f be a function on a subset U of A^n into A^m such that f is differentiable at an interior point u of U. Then by (1.31), (1.33) and (1.27) for each $x = (x_1,\ldots,x_n)$ in A^n

$$Df(u).\, x = \left(\left(\sum_{j \in \underline{n}} D_j(pr_i \circ f)(u).x_j\right); i \in \underline{m}\right)$$

$$= \begin{bmatrix} D_1(pr_1 \circ f)(u) & \cdots & D_n(pr_1 \circ f)(u) \\ \vdots & \ddots & \vdots \\ D_1(pr_m \circ f)(u) & \cdots & D_n(pr_m \circ f)(u) \end{bmatrix} \begin{bmatrix} x_1 \\ \vdots \\ x_n \end{bmatrix}.$$

The matrix $(D_j\ (pr_i \circ f)(u)\ ;\ (i, j) \in m \times n)$ is called the ***derivative matrix*** or ***Jacobian matrix*** of f at u. When m = n, the determinant of the derivative matrix of f at u is called the ***functional determinant*** or ***Jacobian*** of f at u, and is written as Jf(u) or

$$\frac{D(pr_1 \circ f, \ldots, pr_n \circ f)}{D(x_1, \ldots, x_n)}(u).$$

(1.37) Let g be a function on an open suset Z of A^p into an open subset U of A^n and let f be a function on U into A^m such that g is differentiable at a point z in Z and f is differentiable at g(z). Then $f \circ g$ is differentiable at z and

$$(D_k\ (pr_i \circ f \circ g)(z)\ ;\ (i, k) \in \underline{m} \times \underline{p})$$
$$= (D_j(pr_i \circ f(g(z)); (i, j) \in \underline{m} \times \underline{n}).(D_k(pr_j \circ g)(z)\ ;\ (j, k) \in \underline{n} \times \underline{p})$$

(1.38) Let f be a function on an open subset U of a Banach space E over A into a Banach space F over A such that f is differentiable at each point of U. Then f is said to be C^1 or *continuously differentiable* in U iff Df is a continuous function on U into Clm(E, F).

(1.39) Let U (respectively V) be an open convex subset of a Banach space E (respectively F) over A. Let f be a continuous function on $U \times V$ into a Banach space H over A. Then f is c^1 in $U \times V$ iff for each x in U the function $f_x : y \mapsto f(x, y)$ on V into H is c^1 in V, and for each y in V the function

$f_{.y}: x \mapsto f(x, y)$ on U into H is c^1 in U. Moreover for each (u, v) in $U \times V$ and each (x, y) in $E \times F$

$$Df(u, v).(x, y) = D_1 f(u, v).x + D_2 f(u, v).y$$

(1.40) Let f be a continuous bilinear function on the product $E \times F$ of two Banach space E and F over A into a Banach space H over A. Then for each x in X $f_{x.} : y \mapsto f(x, y)$ is a continuos Linear function on F into H and for each y in F $f_{.y}: x \mapsto f(x, y)$ is a continuous linear function on E into H. Hence by (1.28) for each pair of vectors u and x in E and v and y in F,

$$Df_{u.}(v).y = D_2 f(u, v).y = f(u,y)$$

$$Df_{.v}(u).x = D_1 f(u, v).x = f(x, v)$$

and so by (1.39)

$$Df(u,v).(x, y) = f(x, v) + f(u, y).$$

(1.41) Let f be a continuous m-multilinear function on the product of m Banach space $E_1,...,E_m$ over A into a Banach space H over A. Then by (1.40) and induction for each pair of elements $(u_1,...,u_m)$ and $(x_1,...,x_m)$ in

$E_1 \times ... \times E_m$.

$$< (x_1,...,x_m), Df(u_1,...,u_m) > = \sum_{i \in \underline{m}} f(u_1,\ldots,u_{i-1},x_i,u_{i+1},\ldots,u_m)$$

(1.42) Let U, $E_1,...,E_m$ and F be Banach space over A, Let v be any point of an open suhset V of U and for each i in $\underline{m}$ let f_i be a function on V into E_i such that f_i is differentiable at v. Let g be a continuous m-multilinear function on $E_1 \times ... \times E_m$ into F and let h be the funcion

$$V \ni v \mapsto g(f_1(v),...,f_m(v)) \in F$$

Then by (1.29), (1.31) and (1.41) h is differentiable at u and for each u inU

$$Dh(v).(u) = g(Df_1(v).u, f_2(v),...,f_m(v)) + ... + g(f_1(v),..., f_{m-1}(v), Df_m(v).u) .$$

(1.43) Let $E_1,...E_m$, F be Banach space over A, and for each i in $\underline{m}$ let f_i be a function on an open subset U_i of a Banach space over A into E_i such that f_i is differentiable at a point u_i of U_i. Let g be a continuous m-multilinear function on $E_1 \times ... \times E_m$ into F and let h be the function

$$U_1 \times ... \times U_m \ni (w_1,...,w_m) \mapsto g(f_1(w_1),...,f_m(w_m)) \in F$$

Then by (1.29), (1.35) and (1.41) for each point $(w_1,...,w_m)$ in $U_1 \times ... \times U_m$

$$Dh(u_1,...,u_m) . (w_1,...,w_m) = g(Df_1(u_1).w_1, f_2(u_2),...,f_m(u_m)) + ... + g(f_1(u_1),..., f_{m-1}(u_{m-1}), Df_m(u_m).w_m) .$$

(1.44) Let U be an open subset of a Banach space E over A. For each i in $\underline{m}$ let f_i be a C^1 map of U into a Banach space F_i over A. Let g be a C^1 map of U into the Banach space $Cmm(F_1 \times ... \times F_m, H)$ of all continuous m-multilinear maps of $F_1 \times ... \times F_m$ into a Banach space H over A. Let h be the composition of the continuous (m + 1)-multilinear map

$$Cmm(F_1 \times ... \times F_m, H) \times F_1 \times ... \times F_m \ni (p, y_1,...,y_m) \mapsto p(y_1,...,y_m) \in H$$

and the map

$$U \ni x \mapsto (g(x), f_1(x),...,f_m(x)) \in Cmm(F_1 \times ... \times F_m, H) \times F_1 \times ... \times F_m$$

Then for each x in U

$$h(x) = (g(x))(f_1(x),...,f_m(x))$$

Hence by (1.42) for each u in U and each x in E

$$Dh(u).x = (Dg(u).x)(f_1(u),...,f_m(u)) + (g(u))(Df_1(u).x, f_2(u),..., f_m(u)) +... + (g(u))(f_1(u),..., f_{m-1}(u), Df_m(u).x).$$

(1.45) Let f be a function on an open subset U of a Banach space E over A into a Banach space F over A, such that f is two times differentiable at a point u of U. Then for each point (x,y) in E × E

$$D^2f(u).(x, y) = D^2f(u).(y, x)$$

(1.46) ***Definition***

Let $U, V \subset R^n$ be open sets. A function $f : U \to V$ is a diffeomorphism if it is bijective, and if both f and f^{-1} are smooth.

If $f: U \to V$ is a diffeomorphism then the Jacobian matrix Df is non-singular at each point in U.

We now turn to the **Inverse Functon Theorem**.

The Inverse Function Theorem addresses the question of whether a smooth function $f: U \to R^n$ has a smooth inverse (that is, whether it is a diffeomorphism). The one-dimensional case is simple. Let $f: J \to R$ be a smooth function for some open interval J. If $f'(x_0) \neq 0$ for point $x_0 \in J$, then the function is either strictly increasing or strictly decreasing near x_0, and it follows that near x_0 the function has an inverse. Of course, having $f'(x_0) \neq 0$ does not imply that whole function f has an inverse, but only that the function restricted to some (possibly very small) open neighbour-hood of x_0 has an inverse. Since the graph of an inverse function is simply the reflection in the line $y = x$ of original graph, we see that if $f'(x_0) \neq 0$ then the inverse function of f restricted to a neighbourhood of $f(x_0)$ will also be smooth. The Inverse Function Theorem is the higher-dimensional analog of what we have just discussed. The condition $f'(x_0) \neq 0$ is replaced by the condition that the Jacobian matrix has non-zero determinant at the given point.

(1.47) ***Theorem (Inverse Function Theorem)***

Let $U \subset R^n$ be an open set and let $F: U \to R^n$ be a smooth map. If $p \in U$ is a point such that $\det DF(p) \neq 0$ then there is an open set $W \subset U$ containing p such that $F(W)$ is open in R^n and F is a diffeomorphism from W onto $F(W)$.

In Banach space we state the inverse function theorem

(1.48) ***Definition***

Let E and F be Banach space . Let $U \subset E$ and let $f: E \to F$ be of class C^r on E and let $p \in E$. Then f is **locally invertible of class C^r at p** if there exists an open subset U of E with $p \in U$ such that there exists an open set W of F with $f(p) \in W$ and a function $g : W \to U$ of class C^r on W such that $g \circ f$ and $f \circ g$ are the identity maps on U and W respectively. The function g is called a local inverse of f at $f(p)$.

(1.49) ***Inverse Function Theorem***

Let U be an open set in E, let $p \in U$, and let $f: U \to F$ be of class C^r on U. If $Df(p) \in Hom(E, F)$ is an invertible linear transformation, then f is locally invertible of class C^r at p. Furthermore if g is a local inverse of f and $q = f(p)$, then $Dg(q) = [Df(p)]^{-1}$.

Remarks:

(1) When the Banach spaces E and F are finite-dimensional, $Df(p)$ being invertible yields ***dimE = dimF***.

(2) A C^∞ - function can have a continuous inverse without the inverse being differentiable. Thus the function $f: R \to R : x \mapsto x^3$ is of class C^∞ on R and has continuous inverse $g : R \to R : y \mapsto y^{1/3}$. However, g is not differentiable at the point 0.

(3) The various results concerning the inverse function theorem also apply to C^∞ and analytic functions.

The inverse function theorem is useed to obtain the following result.

(1.50) ***Theorem***

Let V_1, V_2, W be real Euclidean vector spaces, let U be open in

$V_1 \times V_2$ and let $f: U \to W$ be of class C^r on U. If for $(a, b) \in U \subset V_1 \times V_2$ the linear transformation $D_2\, f(a, b): V_2 \to W$ is invertible, then $\boldsymbol{dim} V_2 = \boldsymbol{dim} W$ and the map

$$F: U \to V_1 \times W : (x, y) \to (x, f(x, y))$$

is locally invertible of class C^r at (a, b).

(1.51) ***Implicit function Theorem***

Let U be open in $V_1 \times V_2$ and let $f: U \to W$ be a map of class C^r such that for a given $(a, b) \in U$, $f(a, b) = 0$.

If $D_2 f(a, b): V_2 \to W$ is invertible, then there exists an open ball B in V_1 with center at $a \in V_1$, and there exists a uniquely determined map $g: B \to V_2$ of class C^r on B such that $g(a) = b$ and $f(x, g(x)) = 0$ for all $x \in B$.

Proof: For part (a) we note that since $D_2 f(a, b)$ is invertible, $\boldsymbol{dim} V_2 = \boldsymbol{dim} W$. Thus writing F in coordinates, we have $F(x, y) = (F_1(x, y), F_2(x, y))$ where $F_1(x, y) = x$ and $F_2(x, y) = f(x, y)$. Consequently $DF(a, b)$ has a matrix given by

$$\begin{bmatrix} D_1F_1(a,b) & D_2F_1(a,b) \\ D_1F_2(a,b) & D_2F_2(a,b) \end{bmatrix} = \begin{bmatrix} I & 0 \\ D_1f(a,b) & D_2f(a,b) \end{bmatrix}$$

which is invertible. Now apply the inverse function theorem.

For part (b) let

$$h: U \to V_1 \times W : (x, y) \to (x, f(x, y))$$

Then by part (a), h has a local inverse denoted by H. Thus since H is a mapping into $V_1 \times V_2$, it can be represented by coordinate functions $H = (H_1, H_2)$ such that

$$H(x, z) = (H_1(x, z), H_2(x, z)) \equiv (x, H_2(x, z)).$$

Then the desired function is given by $g(x) = H_2(x, 0)$. To see that $y = g(x)$ satisfies the equation, we have for $z = 0$ that

$$(x, 0) = (h \circ H)(x, 0) = h(x, H_2(x, 0)) = h(x, g(x)) = (x, f(x, g(x))).$$

To see that $g(a) = b$, we note $(a, g(a)) = (a, H_2(a, 0)) = (a, H_2(a, f(a, b))) = (H \circ h)(a, b) = (a, b)$

The rest of the proof is easy and can be found in Lang [1968].

Example:

The surface given by $z = f(x, y) = x^2 + y^2 - 1$ defines a circle for those points (x, y) with $f(x, y) = 0$. Choose (a, b) on the circle with $b \neq 0$. Then $\partial f(a, b) / \partial y = 2b \neq 0$ so that we can solve locally for y in terms of x.

(1.52)*Cartesian and curvilinear coordinates*

Let R^n be n- dimensional Euclidean space. Let $e_1,...,e_n$ be n mutally orthogonal unit vectors, then the position vector of a point in R^n relative to fixed orthogonal basis $e_1,...,e_n$ in R^n denoted by $(x^1,...,x^n)$ where $x^1,...,x^n$ are castesian coordinates. The cartesian coordinates in R^n are closely related to the concept of the Euclidean scalar product, a bilinear form which associates with each pair of vectors $\xi, \eta \in R^n$ a real number usually denoted by $<\xi, \eta>$; this operation is symmetric and linear in each argument and the form itself is positive definite. In a cartesian coordinates system we have,

$$<\xi,\eta> = \xi^1\eta^1 + ... + \xi^n\eta^n, \text{ where } \xi = (\xi^1,...,\xi^n),\ \eta = (\eta^1,...,\eta^n).$$

But we know that cartesian coordiantes are not always the most convenient way to solve analyticallly many particular problems. This can be shown by writing the equations of curves on a plane in cartesian coordinates x,y. For simple curves, viz., a circle or ellipse, the analytic expression in cartesian coordinates are simple. The equation of a circle of radius a with centre at point A (h,k) is $(x-h)^2 + (y-k)^2 = a^2$. The equation of an ellipse is $\left(\frac{x-h}{a}\right)^2 + \left(\frac{y-k}{b}\right)^2 = 1$, where a and b are the principal semiaxes. However in various applications and concrete mechanical and physical problems we often deal with smooth curves (also known as trajectories of the motion of a particle in a force field) whose equations in cartesian coordinates are rather complicated.

For example, the equation $\sqrt{(x^2+y^2)} - e^{\lambda \tan^{-1}\frac{y}{x}} = 0$ defines a spiral in cartesian cordinates. Although this equation is rather simple, it could be written in a more simple form in polar coordinates (r, θ) related to the cartesian coordinates by $x = r \cos\theta$, $y = r \sin\theta$. In polar coordinates the equation of spiral becomes $r = e^{\lambda\theta}$, which demonstrate the character of the motion along the trajectory. Solution of particular problems in mechanics and Physics has called for the invention of others curvilinear coordinate system: cylindrical, spherical, and so on.

Consider an arbitary domain in a Euclidean space R^n as we consider it in mathematical analysis, by a domain we mean an arbitary set D in a Euclidean space whose every point P is contained in D together with a ball of sufficiently small radius with centre at P. Consider another Euclidean space denoted by R_1^n to define the coordinates of the point P in the domain D is to associate a set of numbers to the point, called coordinates. Obviously, an arbitary association (i.e., without additional requirements) will not lead to a good result in that such a correspondence may be devoid of actaul sence. For example if we associate the same set of numbers say (0,...,0) with each point P in D, this is a senseless association. Thus the first requirement to the association is that distinct sets of numbers (coordinates) should corresponds to different points of the domain. The example mentioned above that all the points of the domain D has the same "Coordinates" zero does not satisfy this requirements.

Thus our aim is to associate with each point P of a domain D a set of n real numbers. Apparently this operation gives rise to a set of n functions $x^1(p),\ldots,x^1(p)$ defined in the domain D; Here $x^1,\ldots,x^n$ are coordinates in the Euclidean space R_1^n.

Usually these functions $x^1(p),...,x^n(p)$ required to be continuous and even smooth, that is, a small change in the position of P should lead to a small change in its coordinates and a smooth deformation of P should generate a smooth variation of its coordinates.

So, consider two copies of a Euclidean space: R^n with cartesian coordiantes $y^1,...,y^n$ and R_1^n with cartesian coordinates $x^1,...,x^n$; Let D be a domain in R^n.

Definition (1): A continuous coordinate system in a domain D of Euclidean space R^n is said to be a system of functions $x^1(y^1,...,y^n),...,x^n(y^1,...,y^n)$ which map the domain D continuously and bijectively onto a certain domain D_1 of R_1^n. In other words, the system of functions $x^1(p),...,x^n(p)$ defines a mapping, called a homeomorphism of D onto D_1.

In definition (1) we have formally expressed our desire that a point P moves continuously in D, i.e. its coordinates should also change continuously. The functions $x^1(p),...,x^n(p)$ are called the coordinates of point P relative to the coordinate mapping f: D $\to$ D_1.

For example, If the coordinate mapping f: D $\to$ D_1 may be chosen in the form of an identity mapping defined by the linear functions, $x^1 = y^1,...,x^n = y^n$.

Some times we shall also write a point P with coordinates $x^1(p),...,x^n(p)$ in the form $P(x^1,...,x^n)$ assuming that the coordinate mapping f: D $\to$ D_1 has already defined.

Among all continuous coordinate mappings of special interest are those that define a smooth mapping of a domain D onto D_1, i.e., when all functions $x^1(y^1,...,y^n),...,x^n(y^1,...,y^n)$ are continuous functions of their arguments $y^1,...,y^n$. But the smoothness of the coordinate mapping f without the assumption of the smoothness of the inverse mapping f^{-1} does not lead to a meaningful coordinate system. Therefore, we now turn to defining coordinate system in which f and f^{-1} are both smooth. So we shall need a new concept, the Jacobi matrix of a smooth mapping.

Let f: D $\rightarrow$ D_1 be a smooth mapping defined by a set of functions $x^1(y^1,...,y^n),...,x^n(y^1,...,y^n)$

***Definition (2)*:** The Jacobi matrix of a mapping f is a functional matrix

$$df = \frac{\partial x}{\partial y} = \begin{pmatrix} \frac{\partial x^1}{\partial y^1} & \cdots & \frac{\partial x^1}{\partial y^n} \\ \vdots & & \vdots \\ \frac{\partial x^n}{\partial y^1} & \cdots & \frac{\partial x^n}{\partial y^n} \end{pmatrix}$$

composed of partial derivatives of the coordinates $x^1(p),...,x^n(p)$. The determinant of this matrix is denoted by J(f) and called the Jacobian of the mapping f.

***Definition (3)*.** A regular coordinate system in a domain D of Euclidean space R^n is a system of smooth functions $x^1(y^1, ..., y^n)$, ..., $x^n(y^1, ...y^n)$, which map bijectively the domain D onto a domain D_1 in $R_1{}^n$ and are such that the Jacobian J(f)(P) is not zero at all points of D.

Here the condition that the Jacobian does not vanish at all points of D means that the inverse mapping f^{-1} is not only continuous, but also smooth. These follows from the implicit function theorem thus, a regular coordinate system is defined by two smooth mutually inverse mappings establishing a homeomorphism between the domain D and D_1. Definition (3) makes our desire formal that when a point P changes smoothly in D its coordinates should also changes smoothly; moreover, smooth variation of a "coordinate point" B in D_1 should also result in smooth variation of the point p induced by the mapping f^{-1}.

These definitions can be interpreted from another point of view. We could assume that a cartesian coordinate system is initially introduced in a domain D of Euclidean space R^n (through an identity mapping of D onto D_1 under natural identification of both copies, R^n and $R_1{}^n$). Then, the introduction in D of another coordinate system defined by a regular mapping f (i.e. a smooth, one-to-one mapping with a non-zero Jacobian) may be considered as a coordinate transformation we simply pass from the initial cartesian coordinate system to a new one in the same domain D.

***Definition (4)*:** A regular coordinate system in a domain D is some times called a curvilinear system in D.

Let $x^1(p),\ldots,x^n(p)$ and $z^1(p),\ldots,z^n(p)$ are two arbitrary curvilinear coordinate systems in a domain D. This means that there are two mappings $f: D \to D_1 \subset R_1^n(x^1,\ldots,x^n)$ and $g: D \to B \subset R_2^n(z^1,\ldots,z^n)$ which are defined and map smoothly and bijectively the domains D, D_1 and D, B respectively. In other words, each point p in D is associated with two sets of curvilinear coordinates $\{x^i(p)\}$ and $\{z^i(p)\}$, $1 \le i \le n$. Since this correspondence is bijective, we may consider a correspondence $\psi_{x,z}: D_1 \to B$, i.e.,

$$\psi_{x,z}: x^i(p) \to z^i(p), 1 \le i \le n$$

which relates the coordinates $\{x^i(p)\}$ of a point p to the coordinates $\{z^i(p)\}$. The mapping $\psi_{x,z}$ is called coordinate transformation in the domain D. Under this transformation the initial curvilinear coordinates $\{x^i(p)\}$ of a point p change to new curvilinear coordinates $\{z^i(p)\}$. Clearly the transformation $\psi_{x,z}$ is a bijective and smooth mapping of the domain S_1 onto B with a non-zero Jacobian. We can write $\psi_{x,z} = g \circ f^{-1}: D_1 \to B$ and Jacobi matrix of $\psi_{x,z}$ splits into the product of Jacobi matrice of the mapping f^{-1} and g, i.e., $d\psi_{x,z} = \left(\frac{dz}{dx}\right) = dg.(df^{-1})$.

Thus the coordinate system describing the position of a point in space is an indispensable tool for studying geometrical objects. Using coordinate system, we can apply the methods of differential and integral calculus to solve various problems. Therefore, an analysis of space which admit such concepts as differentiable or smooth functions and differentiation and integration has emerged as an independent branch of geometry.

2

Submanifolds and Manifolds

We learn differential and integral calculus in the context of Euclidean space R^n, but there are problems involving "curved" spaces to which necessarily, one want to appply calculus. Geodesy and cartography, are among the few who devoted to the study of the most familiar curved surface of all, the surface of planet Earth. In the study of maps of the Earth, latitude and longitude serve as "coordinates", which allow us to use calculus by considering functions on the Earth's surface (temperature, height above sea level, etc.) as being functions of latitude and longitude. The familiar Mercator's projection, with its stretching of the polar regions, vividly suggests us that these coordinates are badly behaved at the poles : that is, that they are not defined everywhere; they are not "global." (we shall consider such coordinates as being "local", even though they might cover a huge portion of the surface. Of course we may use two sets of "polar" projections to study the Arctic and Antarctic regions. With these three maps we can study the entire surface, provided we know how to relate the Mercator to the polar maps.

The study of smooth surfaces is somewhat like the study of smooth curves. In both cases we define what it means to be smooth via parametrizations, define tangent and normal vectors. There are, however, two fundamental technical differences between surfaces and curves: A surface usually cannot be presented via a single parametrization, and there is no analog for surfaces to unit speed parametrizations.

Submanifolds

(2.1) *Coordinate Patches and Smooth Surfaces*

A topological surface, which is by definition a subset of R^n, need not have a parametrization given as part of its definition. A parametrization for a surface would minimally mean a map from a subset of R^2 onto the surface (or a piece of one since even simple surfaces such as the sphere will need to be parametrized in pieces). To make precise this notion, which will be needed to define smooth surfaces, recall the definition of a topological surface: A subset $Q \subset R^n$ is a topological surface if for each $p \in Q$ there is an open subset $W \subset Q$ containing p such that W is homeomorphic to the open disk ***int***$D^2 \subset R^2$. It would make no difference if the open disk ***int***D^2 were replace by an arbitrary open set $U \subset R^2$,and we will do so for convenience. Now, fix the point $p \in Q$. Let ϕ a choice of homeomorphism $U \to W$. Such a map is the foundation for our parametrization of surfaces. Since an arbitrary continuous map can have a very crinkled image, we add the requirement that ϕ be a smooth function. Even a smooth map can degenerate, however, and in order to avoid this problem we need more conditions on the map ϕ. If $\phi: U \to R^3$ is a smooth map, we can write ϕ in coordinates as

$$\phi(v) = \begin{pmatrix} f_1(v) \\ f_2(v) \\ f_3(v) \end{pmatrix} \text{for all } v \in U$$

where $f_1, f_2, f_3: U \to R$ are smooth real-valued functions. The function ϕ thus has two partial derivatives, namely

$$\phi_1 = \begin{pmatrix} \frac{\partial f_1}{\partial s} \\ \frac{\partial f_2}{\partial s} \\ \frac{\partial f_3}{\partial s} \end{pmatrix} \text{ and } \phi_2 = \begin{pmatrix} \frac{\partial f_1}{\partial t} \\ \frac{\partial f_2}{\partial t} \\ \frac{\partial f_3}{\partial t} \end{pmatrix}$$ observe that ϕ_1 and ϕ_2 are the columns

of the Jacobian matrix $D\phi$. To insure that the map ϕ is not degenerate we make the following definition.

(2.2) *Definition*

Let $U \subset R^2$ be an open set. A smooth map $\phi: U \to R^3$ is a coordinate **patch** if it is injective and if $\phi_1 \times \phi_2 \neq 0$ at all points of U.

Equivalent conditions to the above definition are that ϕ_1 and ϕ_2 are linearly independent, or that the Jacobian matrix $D\phi$ has rank 2 at each point of U.

(2.3) *Examples*:

(1) Let $\phi: R^2 \to R^3$ be given by $\phi\left(\begin{pmatrix} s \\ t \end{pmatrix}\right) = \begin{pmatrix} s \\ t \\ s^2 + t^2 \end{pmatrix}$.

To see that ϕ is injective, observe that $\phi\left(\begin{pmatrix} s \\ t \end{pmatrix}\right) = \phi\left(\begin{pmatrix} u \\ v \end{pmatrix}\right)$

implies $s = u$ and $t = v$. The partial derivatives of ϕ are

$$\phi_1 = \begin{pmatrix} 1 \\ 0 \\ 2s \end{pmatrix} \text{ and } \phi_2 = \begin{pmatrix} 0 \\ 1 \\ 2t \end{pmatrix}.$$

Hence $\phi_1 \times \phi_2 = \begin{pmatrix} -2s \\ -2t \\ 1 \end{pmatrix}$ which is never zero. Therefore ϕ

is a coordinate patch.

(2) Let $\psi: R^2 \to R^3$ be given by $\psi\left(\begin{pmatrix} s \\ t \end{pmatrix}\right) = \begin{pmatrix} s^3 \\ t^3 \\ 1 \end{pmatrix}$

To see that ψ is injective, observe that $\psi\left(\begin{pmatrix} s \\ t \end{pmatrix}\right) = \psi\left(\begin{pmatrix} u \\ v \end{pmatrix}\right)$

implies $s^3 = u^3$ and $t^3 = v^3$, and hence $s = u$ and $t = v$. The partial derivatives of ψ are

$$\psi_1 = \begin{pmatrix} 3s^2 \\ 0 \\ 0 \end{pmatrix} \text{ and } \psi_2 = \begin{pmatrix} 0 \\ 3t^2 \\ 0 \end{pmatrix}$$

Hence

$$\psi_1 \times \psi_2 = \begin{pmatrix} 0 \\ 0 \\ 9s^2t^2 \end{pmatrix}$$

which is zero whenever $s = 0$ or $t = 0$. Therefore ψ is not a coordinate patch.

Coordinate patches now allow us to define smooth surfaces.

(2.4) ***Definition***

A subset $X \subset R^3$ is a **smooth surface** if it is a topological surface and if for each point $p \in X$ there is a cooordinate patch $\phi: U \rightarrow X \subset R^3$ such that $p \in \phi(U)$.

In practice, rather than finding a coordinate patch for each point p in a smooth surface we simply find coordinate patches whose images cover the entire surface. In many cases more than one coordinate patch will be needed.

(2.5) ***Examples:***

(1) Any open subset $U \subset R^2$ is a smooth surface covered by the coordinate patch $\phi: U \rightarrow R^3$ given by

$$\phi\left(\begin{pmatrix} s \\ t \end{pmatrix}\right) = \begin{pmatrix} s \\ t \\ 0 \end{pmatrix}$$

(2) The unit sphere S^2 is a smooth surface . One method to cover S^2 with coordinate patches is to use the six coordinate patches $x^1, x^2, y^1, y^2, z^1, z^2$: $\boldsymbol{intD^2} \rightarrow S^2$ is given by

$$x^1\left(\begin{pmatrix} s \\ t \end{pmatrix}\right) = \begin{pmatrix} \sqrt{1-s^2-t^2} \\ s \\ t \end{pmatrix}, \quad x^2\left(\begin{pmatrix} s \\ t \end{pmatrix}\right) = \begin{pmatrix} -\sqrt{1-s^2-t^2} \\ s \\ t \end{pmatrix}$$

$$y^1\left(\begin{pmatrix} s \\ t \end{pmatrix}\right) = \begin{pmatrix} s \\ \sqrt{1-s^2-t^2} \\ t \end{pmatrix}, \quad y^2\left(\begin{pmatrix} s \\ t \end{pmatrix}\right) = \begin{pmatrix} s \\ -\sqrt{1-s^2-t^2} \\ t \end{pmatrix}$$

$$z^1\left(\begin{pmatrix} s \\ t \end{pmatrix}\right) = \begin{pmatrix} s \\ t \\ \sqrt{1-s^2-t^2} \end{pmatrix}, \quad z^2\left(\begin{pmatrix} s \\ t \end{pmatrix}\right) = \begin{pmatrix} s \\ t \\ -\sqrt{1-s^2-t^2} \end{pmatrix}$$

Each of these coordinate patches covers an open hemisphere:

The space like the surface of the Earth that can be covered by a family of local coordinate systems is defined as a manifold. Thus a manifold will turn out to be the most general space in which one can use differential and integral calculus with roughly the same facility as in euclidean space.

The most familiar manifold is n-dimensional euclidean space R^n, that is, the space of ordered n tuples $(x^1, ..., x^n)$ of real numbers. Here first of all, before defining manifold in general we shall talk about the more familiar concept of a submanifold of R^n, generalizing the notions of curve and surface in R^3.

(2.6) *Submanifolds of Euclidean Space*

When we ask the questions that what is the configuration space of a rigid body fixed at one point of R^N? In answering this question we require submanifolds of euclidean space.

Euclidean space, R^n, with a global coordinate system $(x^1, ..., x^n)$ is the most important example of a manifold.

In familiar R^3 space, with coordinates (x, y, z), a locus z = F (x, y) describes a (2-dimensional) surface, whereas a locus of the form y = G (x), z = H (x), describes a (1-dimensional) curve. In the sequel we need to consider higher-dimensional versions of these important notions.

A subset $X = X^n \subset R^{n+r}$ is said to be an n-dimensional submanifold of R^{n+r}, if locally X can be described by giving r of the coordinates differentiably in terms of the n remaining ones. This means that given $p \in X$, a neighborhood of p on X can be described in some coordinate system $(x, y) = (x^1, ..., x^n, y^1, ...y^r)$ of R^{n+r} by r differentiable functions.

$$y^i = f^i(x^1, ..., x^n),\ i = 1, ...\ r$$

We abbreviate this by $y = f(x)$, or even $y = y(x)$. We say that $x^1, ..., x^n$ are local(curvilinear) coordinates for x near p.

Examples :

(i) $y^1 = f(x^1, ..., x^n)$ describes an n-dimensional submanifold of R^{n+1}.

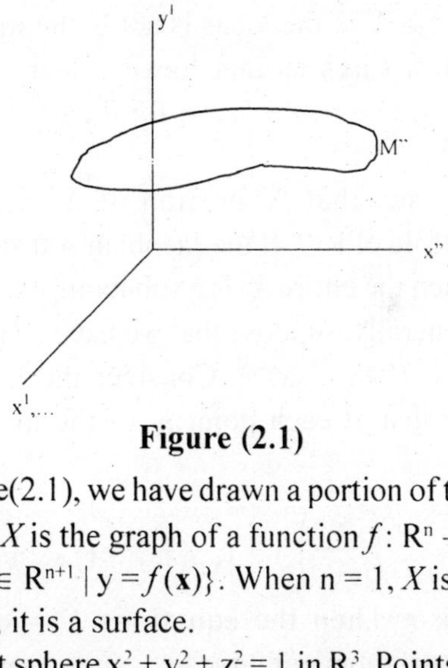

Figure (2.1)

In figure(2.1), we have drawn a portion of the submanifold X. This X is the graph of a function $f: R^n \to R$, that is $X = \{(\mathbf{x}, y) \in R^{n+1} \mid y = f(\mathbf{x})\}$. When n = 1, X is a curve; while if n = 2, it is a surface.

(ii) The unit sphere $x^2 + y^2 + z^2 = 1$ in R^3. Points in the northern hemisphere can be described by $z = F(x, y) = (1 - x^2 - y^2)^{1/2}$ and this function is differentiable everywhere except at the equator $x^2 + y^2 = 1$. Thus x and y are local coordinates for the northern hemisphere except at the equator. For points on the equator one can solve for x or y in terms of the others. If we have solved for x then y and z are the two local coordinates. For points in the southern hemisphere

one can use the negative square root z. The unit sphere in R^3 is a 2-dimensional submanifold of R^3. We note that we have not been able to describe the entire sphere by expressing one of the coordinate, say z, in terms of the two remaining ones, $z = F(x,y)$. We settle for local coordinates. More generally, given r functions $F^i(x^1,...,x^n,y^1,..., y^r)$ of $n + r$ variables, we may consider the locus $X \subset R^{n+r}$ defined by the equations. $F^i(x,y) = c^i$, $(c^1, ..., c^r)$ constants
If the **Jacobian** determinant

$$\frac{\partial(F^1,...,F^r)}{\partial(y^1,...,y^r)}(x_0, y_0)$$

at $(x_0, y_0) \in X$ of the locus is not 0, the **implicit function theorem** assures us that locally, near (x_0, y_0), we may solve $F^i(x, y) = c^i$, $i = 1, ...r$, for the y's in terms of the x's $y^i = f^i(x^1, ..., x^n)$
We may say that "a portion of X^n near (x_0, y_0) is a submanifold of R^{n+r}. If the Jacobian $\neq 0$ at all points of the locus, then the entire X^n is a submanifold.
More generally, supoose that we have r functions of (n+r) variables, $F^i(x^1,...,x^{n+r})$. Consider the locus $F^i(x) = c^i$. Suppose that at each point x_0 of the locus the jacobian matrix.

$$\left(\frac{\partial F^i}{\partial x^j}\right) \quad i = 1, ...r, j = 1 ,..., n + r$$

has rank r. Then the equations $F^i = c^i$ define an n-dimensional submanifold of R^{n+r}, since we may locally solve for r of the coordinates in terms of the remaining n. In geometry we know, two surface F = 0 and G = 0 in R^3 intersect to yield a curve X. The simplest case is one function F of n variables $(x^1, ..., x^n)$. If at each point of the locus F = C there is always at least one partial derivative that does not vanish, then the Jacobian (row) matrix $\left[\partial F/\partial x^1 . \partial F/\partial x^2 ,..., \partial F/\partial x^n\right]$ has rank 1 and we may conclude that this locus is indeed an (n – 1) - dimensional submanifold of R^n. This criterion is easily varified for example, in the case of the 2-sphere F (x, y, z) = $x^2 + y^2 + z^2 = 1$. The column version of this row matrix is called in calculus the gradient vector of F. In R^3 this vector.

$$(\partial F/\partial x, \partial F/\partial y, \partial F/\partial z)^T$$

is orthogonal to the locus F = 0 and we may conclude, for example, that if this gradient vector has a nontrivial component in the z-direction at a point of F = 0, then locally we can solve for $z = z(x, y)$.

A submanifold of dimension (n – 1) in R^n, that is of "**codimension**" 1. is called a hypersurface.

(iii) The x-axis of the *xy* – plane R^2 can be described (perversely) as the locus of the quadratic F $(x, y) = y^2 = 0$. Both partial derivatives vanish on the locus, the x-axis and our criteria would not allow us to say that the x-axis is a 1-dimensional submanifold of R^2. Of course the x-axis is a submanifold, we should have used the usual description G(x, y) = y = 0. Our jacobian criteria are sufficient condition, not necessary ones.

(iv) The locus F $(x, y) = xy = 0$ in R^2, consisting of the union of the *x* and *y* axes. is not a 1-dimensional submanifold of R^2. It seems "clear" (and can be proved) that in a neighbhorhood of the intersection of the two lines we are not going to be able to describe the locus in the form of $y = f(x)$ or x = g(y), where *f*, g are differentiable functions. The best we can say is that this locus with the origin removed is a 1-dimensional submanifold.

Manifolds

The most natural and fundamental concept of differential geometry (and, in fact, much of modern mathematics) is that of *manifold.* We will start with an informal version of the definition, consider two examples, and then show how to turn our ideas into a formal defintion.

(2.7) *Definition (informal)*

A set X is an n-dimensional manifold if it satisfies both:

1. Near each point $x \in X$ there is a good system of n coordinates, and
2. Coordinate systems are compatiable with one another.

Before giving the formal definition (Definition 2.8) we will make precise what 'good' and 'compatible' mean. Intuitively, good coordinates should be well defined functions on the part of the manifold they parametrize, good coordinates should also be appropriate independent variables for differentiation "Compatibility" means that on any overlap the change of coordinates function should be infinitely defferentiable.

For example, the circle is a one-dimensional manifold. Except for one fixable glitch, the usual angle variable is a good coordinate. We will do this example in detail. Consider the circle as the set of $\begin{pmatrix} x \\ y \end{pmatrix}$ in R^2 satisfying $x^2 + y^2 = 1$. i.e., $\left\{\begin{pmatrix} x \\ y \end{pmatrix} \in R^2 \middle| x^2 + y^2 = 1\right\}$.

Thus, each point $P \in S^1$ is uniquely defined by a pair of numbers, cartesian coordinates x and y. For points of a circle, however, there is no need of giving both coordinates. If the coordinate x of a point P is known, the second coordinate y is easily found from equation of the circle as $y = \pm\sqrt{(1 - x^2)}$, i.e., the coordinate y is uniquely(to within a sign) related to the coordinate x. Furthermore, if a point $P_0 = (x_0, y_0)$ does not lie on the x-axis, i.e. $y_0 \neq 0$, there exists a sufficiently small neighbourhood U of a point P_0 such that for all points $P \in U$ the sign of the coordinate y of P is uniquely determined by the sign of the coordinate y_0 of a point P_0. We may therefore say that points of the circle S^1 are described by one numerical parameter, the Cartesian coordinate x. To be more exact, only the points of the upper semicircle, i.e. the points satisfying $y > 0$ are uniquely defined by one numerical parameter x. Similarly, the points of the lower semicircle, i.e. the points satisfying $y < 0$, are also defined by one numerical parameter x. The parameter x varies within the same range, namely the interval (-1,1) of the real axes, both for the upper and lower semicircles. If we need to describe parametrically points of a circle, including "singular" points $P_1 = (1,0)$ and $P_2 = (-1,0)$ we will have to interchange the coordinates x and y and express x in terms of y through the equation of circle. We can parametrize the circle by the function

$$c : R \ni t \mapsto \begin{pmatrix} \cos t \\ \sin t \end{pmatrix} \in S^1 \qquad \text{...(2.7.1)}$$

We cannot use this t as a coordinate because t is not a well defined function on the circle for each point $\binom{x}{y}$ on the circle (such as $\binom{1}{0}$ there are several corresponding values of t (such as $0, \pm 2\pi$, $\pm 4\pi$ ...). We cannot fix the problem simply by restricting. If we try the parameterization

$$\bar{c} : [0,2\pi) \ni t \mapsto \begin{pmatrix} \cos t \\ \sin t \end{pmatrix} \in S^1 \qquad \text{...(2.7.2)}$$

then t is well-defined as a function on the circle but there is a problem with differentiation at the point $\binom{1}{0}$. Because the point $t = 0$ lies on the boundary of the interval [0, 2π], we cannot naively take a derivative with respect to t there. Although there are ways to get around the problem in this case by considering "one sided" derivatives, we want to keep our constructions simple so that we can generalized them. There is no satisfying way to encode the information that points on the circle with t close to 2π are near the point with $t = 0$. So the parametrization $\bar{c}$ is not good enough and we cannot fix it by extending the interval of definition because that would make t multivalued at some points.

Example 1:

Consider the functions $[0,2\pi) \ni t \mapsto \sin t \in R$ and $[0,2\pi) \ni t \mapsto \sin(t/2) \in R$. Check that both functions are differentiable on $(0,2\pi)$ and have one sided derivatives at 0. then, consider the corresponding functions on the circle $S^1 \to R$, defnied by $\binom{x}{y} \to \sin\left(\bar{c}^{-1}\binom{x}{y}\right)$ and on the circle $S^1 \to R$, defined by $\binom{x}{y} \to \sin\left(\frac{1}{2}\bar{c}^{-1}\binom{x}{y}\right)$ and check informally that the first is differentiable at $\binom{1}{0}$ while the second is not.

Solution:

Since both functions can be extended to functions on R whose derivatives are well known, the derivative of sin*t* is cos*t*, and the derivative of $\sin(t/2)$ is $1/2(\cos(t/2))$. We see from these formulas that the one-sided derivatives of these functions at 0 are respectively 1 and ½. It follows that both functions are differentiable on $(0,2\pi)$. To see whether the corresponding functions on the circle are differentiable at the point $\begin{pmatrix}1\\0\end{pmatrix}$ we need to check, if the values of the functions and their derivatives near 2π match the values at 0. Consider the first function, its value at 0 is 0, and its value sin*t* approaches 0 as *t* approaches 2π. Likewise, its one-sided derivative at 0 is 1, and its derivative cos *t* approaches 1 as t approaches 2π. So the first function is differentiable on the circle. Next consider the second function. Its value at 0 is 0, and its value approaches 0 as t approaches 2π. On the other hand, its one-sided derivative at 0 is ½. but its derivative $1/2(\cos(t/2))$ approaches –½ as t approaches 2π. So the second function is not differentiable on the circle, because its derivatives do not match up as one approach the point $\begin{pmatrix}1\\0\end{pmatrix}$ from the two different sides.

Without a suitable definition of differentibility of a function on the circle, we cannot argue formally. At this stage a question naturally arises: can we introduce such a parametrazation of points of a circle S^1 that all the points are uniquely defined by a numerical parameter? The parameter t, equal to the angle between the x-axis and the radius vector ending at the point P, seems to be the most appropriate approximation, though this parameter is not defined uniquely and if we consider the values of t only in some intervel , say, $0 \le t < 2\pi$, the function associating the angular parameter t with a point $P \in S^1$ will suffer a discontinuity at the point $P_0 = (1,0)$.

Thus we arrive at the following assertion, no continuous function exists on a circle S^1 (as a topological space) whose values uniquely describe points of the circle.

In order to have coordinate near each point of the circle we need to use two different coordinate patches. Consider the parameterizations.

$$c_1:(0,2\pi) \ni t_1 \mapsto \begin{pmatrix} \cos t_1 \\ \sin t_1 \end{pmatrix} \in S^1 \qquad \text{...(2.7.3)}$$

and $$c_2:(-\pi,\pi) \ni t_2 \mapsto \begin{pmatrix} \cos t_2 \\ \sin t_2 \end{pmatrix} \in S^1 \qquad \text{...(2.7.4)}$$

Notice that each point on the circle corresponds to at most one value of t_1 and one value of t_2. Because t_1 and t_2 both range over open intervals, each is a good coordinate. We must also check the overlap condition, at every point $\binom{x}{y}$ on the circle with $y \neq 0$, both t_1 and t_2 are well-defined. On the top half of the circle $t_1 = t_2$, so the change-of-coordinates in either direction is the identity function, which is infinitely differentiable. On the bottom half, $t_1 = t_2 + 2\pi$, so the change-of-coordinates functions are $t_2 \to t_2 + 2\pi$, and $t_1 \to t_1 - 2\pi$, both of which are infinitely differentiable, so the circle S^1 is a one-manifold.

Another important example is the sphere

$$S^2 = \left\{ \begin{pmatrix} x \\ y \\ z \end{pmatrix} \middle| x^2 + y^2 + z^2 = 1 \right\},$$

Just like in the case of a circle, it is unnecessary to give all three cartesian coordinates (x,y,z) for points of the sphere S^2, since one coordinate, say z, can be expressed in terms of the other two, $z = \pm\sqrt{(1-x^2-y^2)}$. Obviously, on the upper (and similarly the lower) hemisphere the coordinate z of a point P is uniquely related to the coordinates x and y.

As shown in the case of the circle, here it can also be shown that there exists no two continuous functions on the two-dimensional sphere S^2 such that there values uniquely defined a point on the sphere. These examples shows that the possibility of the construction of a unique coordinate system for all points of a space under consideration is to be rejected and we have to use different coordinate systems for different parts of a space, a rigorous analysis of the situation has led in geometry to a specific concepts of a manifold.

We show that it is a two-manifold. We can parametrize the unit sphere (many times over) with the usual spherical coordinates,

$$c : R^2 \ni \begin{pmatrix} \theta \\ \phi \end{pmatrix} \mapsto \begin{pmatrix} \sin\theta \cos\phi \\ \sin\theta \sin\phi \\ \cos\theta \end{pmatrix} \in S^2$$

Here we have used the physicists convention : ϕ is the longitude (curves of constant ϕ correspond to meridians from one pole to the other) and θ is the colatitude (curves of constant θ correspond to circles parallel to the equatorial circle). The mathematicians convention reverses the roles of ϕ and θ and has the virute that the θ of spherical coordinates corresponds to the θ of cylindrical coordinates.

As we have been in the circle example, a good parametrization, while helpful, is not necessarily complete the whole story. To build a good set of coordinate systems from a parametrization we must fix redundancies. As before, we can get one good coordinate system by restricting the domain of the parametrization,

$$c_1 : (0,\pi)\times(-\pi,\pi) \ni \begin{pmatrix} \theta_1 \\ \phi_1 \end{pmatrix} \mapsto \begin{pmatrix} \sin\theta_1 \cos\phi_1 \\ \sin\theta_1 \sin\phi_1 \\ \cos\theta_1 \end{pmatrix} \in S^2$$

we can check that θ_1 and ϕ_1 are well-defined functions on most of the sphere.

This coordinate system omits only the north pole ($\theta = 0$), the south pole ($\theta = \pi$) and the international date line ($\phi = \pi$). We can take care of the international date line by the trick we used in the circle example.

$$c_2 : (0,\pi)\times(0,2\pi) \ni \begin{pmatrix} \theta_2 \\ \phi_2 \end{pmatrix} \mapsto \begin{pmatrix} \sin\theta_2 \cos\phi_2 \\ \sin\theta_2 \sin\phi_2 \\ \cos\theta_2 \end{pmatrix} \in S^2$$

It is easy to check that this coordinate system is compatible with the previous one, the change of coordinates on each overlap is infinitely differentiable.

We still do not have good coordinates near the north and south poles. Notice that the failure of spherical coordinates is dramatic : at each pole, a whole continum of values of ϕ correspond to one point. Still, the fix is easy. Near each pole we can use xy-coordinates. Specifically, let D denote the open unit disk in the xy-plane, that is,

$$D = \left\{ \begin{pmatrix} x \\ y \end{pmatrix} \in R^2 : x^2 + y^2 < 1 \right\}$$

and define coordinate systems

$$c_3 : D \ni \begin{pmatrix} x_1 \\ y_1 \end{pmatrix} \mapsto \begin{pmatrix} x_1 \\ y_1 \\ \sqrt{1 - x_1^2 - y_1^2} \end{pmatrix} \in S^2$$

near the north pole and

$$c_4 : D \ni \begin{pmatrix} x_2 \\ y_2 \end{pmatrix} \mapsto \begin{pmatrix} x_2 \\ y_2 \\ \sqrt{1 - x_2^2 - y_2^2} \end{pmatrix} \in S^2$$

near the south pole. The four coordinate systems taken together suffice for the proof that the sphere is a two - manifold.

Example 2:

To prove that the coordinate system defined by c_3 is compatible with the coordinate system defined by c_1.

Solution:

We must look at the change of coordinates function on the overlap and argue that this function and its inverse are infinitely differentiable. First we find the overlap the range of c_1 includes everything but the north and south poles and the international date line. The range of c_3 is the whole northern hemisphere, not including the equator. It follows that the domain of the change of coordinates function $c_3^{-1} \circ c_1$ is $(0, \pi/2) \times (-\pi, \pi)$. In addition, we can calculate that

$$c_3^{-1} \circ c_1 \begin{pmatrix} \theta_1 \\ \phi_1 \end{pmatrix} = \begin{pmatrix} \sin\theta_1 \cos\phi_1 \\ \sin\theta_1 \sin\phi_1 \end{pmatrix}$$

which is infinitely differentiable. Explicity, the Jacobian matrix of this function is

$$\begin{pmatrix} \cos\theta_1 \cos\phi_1 & -\sin\theta_1 \sin\phi_1 \\ \cos\theta_1 \sin\phi_1 & -\sin\theta_1 \cos\phi_1 \end{pmatrix}$$

each of whose entries is infinitely differentiable.

To see that the inverse is infinitely differentable we use the inverse function theorem, which guarantees that wherever the Jacobian matrix of the function is nonsingular, the Jacobian matrix of the inverse equals the inverse matrix of the Jacobian. in our case the determinant of the jacobian is $\cos\theta_1 \sin\theta_1$, which is infinitely differentiable and nonzero everywhere on the domain since $\theta_1 \in (0, \pi/2)$. This, along with Cramer's rule, guarantees that the Jacobian of the inverse has infinitely differentiable entries. Hence the inverse is infinitely differentiable.

We notice that the domains of definition for c_1, c_2, c_3 and c_4 are open sets in R^2. i.e., for each element $(x, y)^T$ of each domain there is a strictly positive number r such that the ball of radius r around $(x, y)^T$,

$$B\left(r, \begin{pmatrix} x \\ y \end{pmatrix}\right) := \left\{\begin{pmatrix} u \\ v \end{pmatrix} \in R^2 : (u-x)^2 + (v-y)^2 < r\right\}$$

lies entirely in the given domain.

Intuitively, every point of an open set lies fully inside the set, so if the domain of the parametrization is open, the coordinates will lend themselves well to differentiation. The problem of differentation that arose with the parametrization $\bar{c}$ in the analysis of the circle was possible only because the domain $[0, 2\pi)$ was not an open set.

In the formal definition we replace the vague notion of a "good system of n coordinates" with the precise concept of an injective function whose domain is an open subset of R^n. Our definition is very close to that of Warner [Definition 2.1.2].

(2.8) *Definition (Formal)*

An n-dimensional manifold is a set X and a family $\{\theta_1, \theta_2 \ldots\}$ of injective functions from domains in R^n into X such that

1. For each i the domain U_i of the function θ_i is an open subset of R^n. The domains U_i and the functions θ_i are called coordinate patches.
2. The union of the coordinate patches cover X. That is $\theta_1(U_1) \cup \theta_2(U_2) \cup \ldots = X$.
3. The coordinate system are compatible on overlaps. That is, if we consider the overlaps $W_{ij} = \theta_i(U_i) \cap \theta_j(U_j)$ then

the sets $\theta_i^{-1}(W_{ij})$ and $\theta_j^{-1}(W_{ij})$ must be open in R^n and the transition functions $F_{ji} = \theta_j^{-1} \circ \theta_i : \theta_i^{-1}(W_{ij}) \to \theta_j^{-1}(W_{ij})$ must be infinitely differentiable.

Lets us review our example of the circle S^1 in light of this definition. The domains in R are $U_1 = (0, 2\pi)$ and $U_2 = (-\pi, \pi)$; these are open, satisfying condition 1 of the definition. The corresponding coorrdinate patches on the circle itself are $c_1(U_1) = S^1 \sim \left\{\begin{pmatrix}1\\0\end{pmatrix}\right\}$ and $c_2(U_2) = S^1 \sim \left\{\begin{pmatrix}-1\\0\end{pmatrix}\right\}$. The union of these two sets is S^1 itself, so condition 2 is satisfied. Finally, we must check condition 3 for the overlap $W_{21} = S^1 \sim \left\{\begin{pmatrix}1\\0\end{pmatrix}, \begin{pmatrix}-1\\0\end{pmatrix}\right\}$. First we note that $c_1^{-1}(W_{21}) = (0, \pi) \cup (\pi, 2\pi)$ which is open in R similarly $c_2^{-1}(W_{21})$ is open in R. Next the transition function $F_{12} : (-\pi, 0) \cup (0, \pi) \to (0, \pi) \cup (\pi, 2\pi)$ can be written explicitly as

$$F_{12} : t_2 \to \begin{cases} t_2 + 2\pi & \text{if } t_2 \in (-\pi, 0) \\ t_2 & \text{if } t_2 \in (0, \pi) \end{cases}$$

which is infinitely differentiable at every point where it is defined. (Note that the jump at $t_2 = 0$ is not a problem because 0 is not in the domain of the function.) Similarly $c_2^{-1} \circ c_1$ is infinitely differentiable on its domain. So the circle S^1 satisfies our formal definition of a one-manifold. A complete treatment of manifolds should introduce the notion of diffeomorphisms between manifolds. While these concepts are crucial to a deep mathematical understanding of differential geometry, they are rarely relevant in classical mechanics. Warner's introduction to manifols deals with these issues openly. For detailed discussion of the underlying topological concepts, the reader might consult Kelley (K).

Example 3:

Show that the second function defined in example 1 is not infinitely differentiable.

Solution:

Consider the function $f : S^1 \to R$ defined by $(x, y)^T \to \sin((1/2)\bar{c}^{-1}(x, y)^T)$, where $c : [0, 2\pi) \to S^1$ is defined by $t \mapsto (\cos t, \sin t)^T$, using the definition in the context of c_2,

$$f \circ c_2(t) = \sin\left(\tfrac{1}{2}c^{-1}(\cos t, \sin t)\right) = \begin{cases} \sin(\tfrac{1}{2}(t + 2\pi)), & t \in (-\pi, 0) \\ \sin(\tfrac{1}{2}(t + 2\pi)), & t \in (0, \pi) \end{cases}$$

Note that the function $f \circ c_2$ is not infinitely differentiable at t = 0 since

$$\lim_{t \to 0^-} \frac{f \circ c_2(t) - f \circ c_2(0)}{t} = -\frac{1}{2} \neq \frac{1}{2} = \lim_{t \to 0^+} \frac{f \circ c_2(t) - f \circ c_2(0)}{t}$$

(2.9) *Definition*

A topological manifold of dimension n is a second countable, Hausdorff topological space X with the property that, for each $x \in X$, there exists an open set U in X containing x and a homeomorphism ϕ of U onto an open subset $\phi(U)$ in R^n. The pair (U, ϕ) is called a chart at x. If (U_1, ϕ_1) and (U_2, ϕ_2) are two charts with $U_1 \cap U_2 \neq \phi$, then the overlap functions $\phi_1 \circ \phi_2^{-1} : \phi_2(U_1 \cap U_2) \to \phi_1(U_1 \cap U_2)$ and $\phi_2 \circ \phi_1^{-1} : \phi_1(U_1 \cap U_2) \to \phi_2(U_1 \cap U_2)$ are inverse homeomorphisms. Let k be any positive integer or ∞. Two charts (U_1, ϕ_1) and (U_2, ϕ_2) are c^k or c^∞—related if either $U_1 \cap U_2 = \phi$ or if $U_1 \cap U_2 \neq \phi$ then the overlap functions are c^k or c^∞. (i.e., have continuous partial derivatives of order k or of all orders and type). An atlas for X is a collection $\{(U_\alpha, \phi_\alpha)\}_{\alpha \in \wedge}$ of charts, any two of which are c^∞ related and with $\bigcup_{\alpha \in \wedge} U_\alpha = X$. If $\phi_1 \circ \phi_2^{-1}$ and $\phi_2 \circ \phi_1^{-1}$ are real analytic, X is then called ananlytic manifod. A chart is admissible to an atlas if it is C^∞—related to every chart in the atlas and the atlas is said to be maximal if it contains every chart that is admissible to it. Every atlas is contained in a unique maximal atlas. A maximal atlas is called a differentiable structure for X. A topological manifold together with some differentiable structure is called a differentiable (or smooth, or C^∞) manifold. The integer n is called the dimension of X and is denoted by dim X.

Some topological manifolds admit no differentiable structures at all, while others admit many.

A c^k manifold X modelled on $\underline{R}^n$ (respectively $\underline{C}^n$) is called a n-dimensional real (respectively complex) c^k- manifold, n being a natural number. A one - dimensional real c^k manifold is called a c^k-curve. A one-dimensional complex c^1 - manifold is called a Riemann surface.

Let (U,ϕ) be a chart for an n-dimensional real or complex c^k - manifold, n being a positive integer. Then for each i in $\underline{n}$, $pr_i \circ \phi$ is called the ith coordinate function induced by ϕ and in denoted by x^i; evidently $\phi = (x^i; i \in \underline{n})$. For each p in $Dm(\phi)$ and each i in $\underline{n}$ $x^i(p)$ is called the ith coordinate of ϕ (p) and so $\phi(p) = \left(x^1(p),\ldots,x^n(p)\right)$ it is denoted by p^i by abuse of notation, evidently, $\phi(p) = \left(x^i(p); i \in \underline{n}\right) = \left(p^i; i \in \underline{n}\right)$

$\left(x^1(p),\ldots,x^n(p)\right)$ is called the set of local coordinates of the point p of U with respect to the coordinate neighbourhood (U,ϕ) and the n-tuple $\left(x^1,\ldots,x^n\right)$ of functions on U is called the local coordinate system on (U,ϕ).

Let n be any element of N. Any complex vector space of dimension n can be considered as a real vector space of dimension $2n$. Hence any compelx $c^k\left(k \in N\right)$ manifold modelled on $\underline{c}^n$ can be considered as a real c^k manifold modelled on $\underline{R}^{2n}$.

Let X be a complex c^k manifold modelled on $\underline{C}^n$. Let ϕ be a chart at a point p on X. Then $\phi(p) = \left(x^i(p); i \in \underline{n}\right)$ and when X is considered as a real c^k manifold modelled on $\underline{R}^{2n}$ we write,

$\phi(p) = \left(\left(R\ x^i\right)(p), \left(I\ x^i\right)(p); i \in \underline{n}\right)$ by abuse of notation, and call $\left(R\ x^i\right)(p)$ (respectively $\left(I\,\phi^i\right)(p)$ the ith real (respectively imaginary) coordinate of p relative to ϕ.

2.10 Examples of Manifolds

1. (Standard differentiable structure on $\underline{R}^n$). The Euclidean space $\underline{R}^n$ with the atlas consisting of a single chart $\left(\underline{R}^n, id_{\underline{R}^n}\right)$ where $id_{\underline{R}^n}; \underline{R}^n \to \underline{R}^n$ is an identity map, forms a c^∞ manifold. A chart (U, ϕ) is in the differentiable structure determined by this atlas if and only if $\phi: U \to \phi(U)$ is a diffeomophism.
2. (A non standard differentiable structure on $\underline{R}$). The set $X = \underline{R}$ with the atlas consisting of a single global chart $(\underline{R}, \phi)$, where $\phi: \underline{R} \to \underline{R}$ is given by $\phi(p) = p^3$ is a C^1 - manifold. The resulting differentiable structure on $\underline{R}$ is not the same as the standard structure because $\phi^{-1}: \underline{R} \to \underline{R}$ given by $\phi^{-1}(x) = \sqrt[3]{x}$ is not c^∞ on $\underline{R}$. (in the standard sense). These two manifolds (standard and non standard) have different differentiable structure, but are, nevertheless, diffeomorphic.
3. Let S^1 be the circle in the xy- plane $\underline{R}^2$ centered at the origin and of radius 1 (Fig.(2.2)).

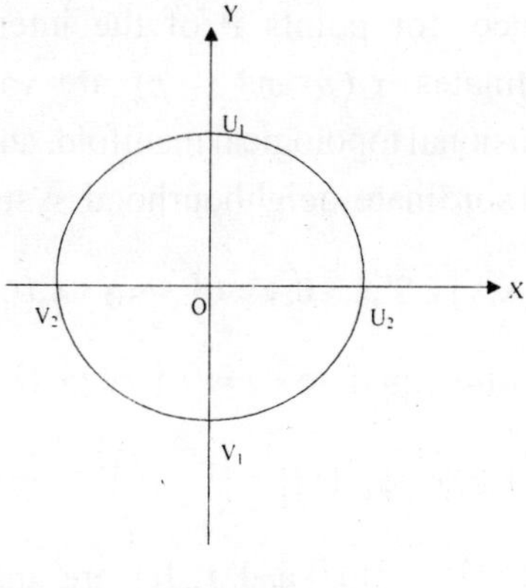

Fig. 2.2

Give S^1 the subspace topology. Let

$$U_1 = \{p = (x, y) \in S^1 | y > 0\},\ V_1 = \{p = (x, y) \in S^1 | y < 0\}$$

$$U_2 = \{p = (x, y) \in S^1 | x > 0\},\ V_2 = \{p = (x, y) \in S^1 | x < 0\}$$

Then U_1 and V_1 are open subsets of S^1, and $U_1 \cup U_2 \cup V_1 \cup V_2 = S^1$. Let ϕ_i and θ_i be the maps from U_i and V_i, respectively, to the open interval $I = \{t | -1 < t < 1\}$, givenby $\phi_1(x, y) = x, \theta_1(x, y) = x$, $\phi_2(x, y) = y, \theta_2(x, y) = y$. The maps ϕ_i and θ_i are homeomorphisms. Thus S^1 is a 1-dimensional topological manifold, in order to prove that the mapping ϕ_i and θ_i, are homeomorphisms for i=1,2 , it is sufficient to write explicitly the inverse mappings

$$\varphi_1^{-1}(x) = \left(x,\ \sqrt{(1 - x^2)}\right) \in S^1,\ \theta_1^{-1}(x) = \left(x,\ -\sqrt{(1 - x^2)}\right) \in S^1$$

$$\varphi_2^{-1}(y) = \left(\sqrt{(1 - y^2)},\ y\right) \in S^1,\ \theta_2^{-1}(y) = \left(-\sqrt{(1 - y^2)},\ y\right) \in S^1$$

and show that they are continuous. Then we obtained on the circle four local coordinate systems, each consisting only of one coordinate $x_1 = \varphi_1(x, y) = x,\ \ x_2 = \theta_1(x, y) = x$ and $x_3 = \varphi_2(x, y) = y,\ \ x_4 = \theta_2(x, y) = y$. Certain points are provided with two local coordinate systems. For instance, for points P of the intersection $U_1 \cap U_2$ the coordinates $x_1(p)$ and $x_3(p)$ are valid. Thus S^1 is a 1-dimensional topological manifold, and $S = \{(U_i, \phi_i), (V_i, \theta_i)\}_{i=1,2}$ is a coordinate neighbourhood system. Now $U_1 \cap U_2 =$

$\{p = (x, y) \in S^1 | x > 0, y > 0\}$, so $\phi_1(U_1 \cap U_2)$, $\phi_2(U_1 \cap U_2)$

are both equal to $J = \{t \mid 0 < t < 1\}$. For $t \in J$, we have

$f_{21}(t) = \phi_2\left(\phi_1^{-1}(t)\right) = \left(1 - t^2\right)^{\frac{1}{2}}$ and $f_{12}(t) = \phi_1\left(\phi_2^{-1}(t)\right)$

$= \left(1 - t^2\right)^{\frac{1}{2}}$, $f_{21}(t)$ and $f_{12}(t)$ are analytic for $0 < t < 1$.

Similarly the other cordinate transformations are also analytic. Thus S^1 is an analytic manifold.

4. **The Unit Sphere:**

$$S^2 = \{(x, y, z) \mid x^2 + y^2 + z^2 = 1\}$$

is a good example of a non trivial two dimensional manifold realized as a surface in R^3.

Let $U_1 = S^2 \sim \{(0,0,1)\}$, $U_2 = S^2 \sim \{(0,0,-1)\}$ be the subset obtained by deleting the north and south poles respectively. Let $\theta_i : U_1 \to R^2 = \{(x, y, 0)\}_{i=1,2}$ be stereographic projections from the respective poles, so $\theta_1(x, y, z) = \left(\frac{x}{1-z}, \frac{y}{1-z}\right)$,

$\theta_2(x, y, z) = \left(\frac{x}{1+z}, \frac{y}{1+z}\right)$. It can be easily checked that on the overlap $U_1 \cap U_2, \theta_1 \circ \theta_2^{-1} : \underline{R}^2 \sim \{0\} \to \underline{R}^2 \sim \{0\}$, is a smooth diffeomorphism, given by the inversion

$$\theta_1 \circ \theta_2^{-1}(x, y) = \left(\frac{x}{x^2 + y^2}, \frac{y}{x^2 + y^2}\right)$$

The Hausdorff separation property follows easily from that of R^3, so S^2 is a smooth, 2 - dimensional manifold. The unit sphere is a particular case of the general concept of a surface in $\underline{R}^3$, which historically provided the principal motivating example for the development of the general theory of manifolds.

5. (Standard differentiable structure on the n - sphere S^n)

Let $X = S^n = \{(x^1, \ldots, x^n, x^{n+1}) \in \underline{R}^{n+1} \mid (x^1)^2 + \ldots + (x^{n+1})^2 = 1\}$

with its topology as a subspace of R^{n+1}. We define two stereographic porjection charts on S^n.

Let $N = (0,...,0,1)$, $S = (0,..., 0, -1)$, $U_S = S^n \sim \{N\}$ and $U_N = S^n \sim \{S\}$. Now define two maps $\phi_S : U_S \to \underline{R}^n$ and $\phi_N : U_N \to \underline{R}^n$ by $\phi_S\left(x^1,...,x^n,x^{n+1}\right) = \left(1 - x^{n+1}\right)^{-1}\left(x^1,...,x^n\right)$ and $\phi_N\left(x^1,...,x^n,x^{n+1}\right) = \left(1 + x^{n+1}\right)^{-1}\left(x^1,...,x^n\right)$. Then $\phi_S^{-1} : \underline{R}^n \to U_S$ and $\phi_N^{-1} : \underline{R}^n \to U_N$ are given by

$$\phi_S^{-1}(y) = \phi_S^{-1}\left(y^1,...,y^n\right) = \left(1+\|y\|^2\right)^{-1}\left(2y^1,...,2y^n,\|y\|^2-1\right)$$

and $\phi_N^{-1}(y) = \phi_N^{-1}\left(y^1,...,y^n\right) = \left(1+\|y\|^2\right)^{-1}\left(2y^1,...,2y^n,-\|y\|^2+1\right)$.

Thus, on $\phi_N(U_N \cap U_S) = \phi_S(U_N \cap U_S) = \underline{R}^n \sim \{0\}$,

$\phi_S \circ \phi_N^{-1}(y) = \phi_N \circ \phi_S^{-1}(y) = \|y\|^{-2}\, y$ and these are c^∞. Thus

$\{(U_S,\phi_S),(U_N,\phi_N)\}$ is an atlas for S^n and so determines a differentiable structure for S^n.

6. Let us consider a projective plane RP^2. By this plane we mean a space, the points of which are straight lines through the origin in R^3. Define the distance between two straight lines as the least angle between them. Then RP^2 becomes a metric space. We now prove that RP^2 is a two-dimensional manifold. To this end, it is convenient to describe any straight line $P \in RP^2$ by three homogeneous coordinates (x : y : z) which admit multiplication by a number $\lambda \neq 0$. Homogeneous coordinates do not vanish simultaneously, i.e. $x^2 + y^2 + z^2 > 0$. Cover RP^2 with three charts: $U_1 = \{(x : y : z) : x \neq 0\}$, $U_2 = \{(x : y : z) : y \neq 0\}$ and $U_3 = \{(x : y : z) : z \neq 0\}$. Let $V_1 = V_2 = V_3 = R^2$. The mappings $\theta_i : U_i \to V_i = R^2$ are taken as coordinate homeomorphism, namely, $\theta_1(x : y : z) = \left(\frac{y}{x}, \frac{z}{x}\right)$, $\theta_2(x : y : z) = \left(\frac{x}{y}, \frac{z}{y}\right)$,

$\theta_3(x : y : z) = \left(\frac{x}{z}, \frac{y}{z}\right)$. Thus we have constructed three local coordinate systems

$x_1^1 = \frac{y}{x},\ x_1^2 = \frac{z}{x};\ x_2^1 = \frac{x}{y},\ x_2^2 = \frac{z}{y};\ x_3^1 = \frac{x}{z},\ x_3^2 = \frac{y}{z};$.

To verify that the mappings θ_i arc homeomorphisms it is sufficient to construct the inverse mappings

$\theta_1^{-1}(x_1^1, x_1^2) = (1 : x_1^1 : x_1^2)$, $\theta_2^{-1}(x_2^1, x_2^2) = (x_2^1 : 1 : x_2^2)$

and $\theta_3^{-1}(x_3^1, x_3^2) = (x_3^1 : x_3^2 : 1)$ and prove that they are continuous.

7. Natural differentiable structure on real vetor space) Let $X = \upsilon$ be a real vector space of dimension n. Select a basis $\{e_1,...,e_n\}$ for υ and Let $\{e^1,...,e^n\}$ be the dual basis for υ^*. Define $\phi : \upsilon \to \underline{R}^n$ as follows. For each $v = v^i e_i \in \upsilon$, $\phi(v) = \left(e^1(v),...,e^n(v)\right) = \left(v^1,...,v^n\right)$. Then ϕ is an isomorphism. Provide υ with the topology for which ϕ is a homeomorphism (this is independent of the choice of basis). Now provide υ with the differentiable structure determined by the single global chart (υ, ϕ) (also independent of the choice of basis), when $\upsilon = R^n$ this reduces to the standard differentiable structure.

3

Differentiable Functions and Morphisms of Manifolds

Differentiable Functions

The first big consequence from the definition of a manifold is that we can define the notion of a differentiable function from one manifold to another. To test differentiability of a function, we express the function in terms of coordinates on the range and domain and test differentiability of the coordinate expressions. We will start with an informal definition (Definition 3.1) and do a few examples in detail before giving a formal definition (Definition 3.2).

(3.1) ***Definition (informal)***

A function f from a manifold X to a manifold Y is infinitely differentiable if it is a infinitely differentiable in all coordinate charts.

Example 1:

Consider a function from the circle to the real line. We can ask whether the function $f\colon S^1 \to R$, defined by $(x, y) \mapsto y$ is infinitely differentiable. The essence of the calculation is to write this function in each coordinate chart. Here we have two coordinate chart on the domain (see equations 2.7.3 and 2.7.4) and one coordinate chart on the range (since the identity function suffices to prove that R is a one-manifold). Writing *f* in t_1-coordinates we have $f \circ c_1 : (0, 2\pi) \to R$ defined by $t_1 \mapsto \sin t_1$ and $f \circ c_2 : (-\pi, \pi) \to R$, defined by $t_2 \mapsto \sin t_2$, because these functions are both infinitely differentiable, *f* is infinitely differentiable.

Example 2:

Consider a function which takes values in a nontrivial manifold. We will define a function from the unit circle onto the 40th north parallel on the unit sphere. (On the earth, the 40th north parallel passes through Salt Lake City, Philadephia. Madrid, Mt. Olympus, Ankara and Beijing). Specifically, consider.g : $S^1 \to S^2$

$$\begin{pmatrix} x \\ y \end{pmatrix} \mapsto \begin{pmatrix} x \sin\frac{5\pi}{18} \\ y \sin\frac{5\pi}{18} \\ \cos\frac{5\pi}{18} \end{pmatrix}$$

Since we have two coordinate chart on S^1 and four on S^2, we have $2 \times 4 = 8$ coordinate versions of the function g. Fortunately we do not have to check all eight explicitly. Two are completely irrelevant $-x_2, y_2$ coordinates as a function of t_1 or of t_2—because $\cos 5\pi / 18 > 0$ and so no value of g lies in the portion of the sphere covered by x_2, y_2 coordinates. Of the six remaining ways to write g in coordinates we can get away with checking only two : x_1, y_1 coordinates as a functicn of t_1 and x_1, y_1 coordinates as a functior of t_2. Because these two coordinate versions describe g completely and because the coordinate systems are compatible, differentiabillity in other coordinates will follow from differentiability in these.

Writing x_1, y_1 - coordinates as a function of t_1 using g, we have $(0, 2\pi) \to D$

$$t_1 \mapsto \begin{pmatrix} \cos t_1 \sin\frac{5\pi}{18} \\ \sin t_1 \sin\frac{5\pi}{18} \end{pmatrix} \quad ...(5)$$

which is infinitely differentiable. Likewise, $(-\pi, \pi) \to D$

$$t_2 \mapsto \begin{pmatrix} \cos t_2 \sin\frac{5\pi}{18} \\ \sin t_2 \sin\frac{5\pi}{18} \end{pmatrix}$$

is infinitely differentiable. We conclude that the function $g : S^1 \to S^2$ is infinitely differentiable.

With these examples in our mind let us consider the formal definition.

(3.2) *Definition (Formal)*

Given a manifold X, a manifold Y and a function f: X → Y, we say that f is infinitely differentiable if for every coordinate system $\theta_i : U_i \to X$ and every coordinate system $\phi_j : V_j \to Y$, the function $\phi_j^{-1} \circ f \circ \theta_i : U_i \to V_j$ is infinitely differentiable.

(3.3) *Morphism of Manifolds (Definition)*

Let X and Y be smooth manifolds of dimension m and n respectively, and let $f: X \to Y$ be a continuous map. Let (U,θ) be a chart on X and (V,ϕ) be a chart on Y such that $f(U) \subset V$. Then the function $\phi \circ f \circ \theta^{-1}$ of $\theta(U)$ into $\phi(V)$ is called the coordinate expression for f or local respresentative of f relative to these two charts and is denoted by $f_{\theta\phi}$, we say that f is differentiable at p of class C^∞ or C^∞ manifold morphism of X into Y at a point $p \in X$ if there exist a coordinate system (U,θ) at p in X and a coordinate system (V,ϕ) at $f(p)$ in Y such that $\phi \circ f \circ \theta^{-1} : \theta(U) \to \phi(V)$ is differentiable at $\theta(p)$ of class C^∞. Note that $\theta(U) \subset R^m$ and $\phi(V) \subset R^n$. Since differentiability is given in terms of specific charts, we must show that it is actually independent of the choice of charts. Thus let $(\bar{U},\bar{\theta})$ (respectively $(\bar{V},\bar{\phi})$) be any other member of the atlas for

X (respectively Y) which are neighbourhood of p (respectively $f(p)$). Then we must show that the map

$$f_{\overline{\theta}\overline{\phi}} = \overline{\phi} \circ f \circ \overline{\theta}^{-1} : \overline{\theta}(\overline{U}) \to \overline{\phi}(\overline{V})$$

is differentiable at $\overline{\theta}(p)$. However, since differentiabiity is a local property, it suffices to show this on neighbourhood. Thus for $U \cap \overline{U}$ and $V \cap \overline{V}$ (which are non empty), we have on the neighbourhoods $\overline{\theta}(U \cap \overline{U})$ and $\overline{\phi}(V \cap \overline{V})$ that

$$\overline{\phi} \circ f \circ \overline{\theta}^{-1} = \overline{\phi} \circ \phi^{-1} \circ \phi \circ f \circ \theta^{-1} \circ \theta \circ \overline{\theta}^{-1} = \overline{\phi} \circ \phi^{-1} \circ (f_{\theta\phi}) \circ \theta \circ \overline{\theta}^{-1}.$$

Thus since $\overline{\phi} \circ \phi^{-1}, \phi \circ f \circ \theta^{-1}$ and $\theta \circ \overline{\theta}^{-1}$ are of class C^{∞}, so is their composition. If $f: X \to Y$ is C^{∞} differentiable at every point $p \in X$, then f is differentiable map of class C^{∞} from X into Y.

Now in terms of coordinates, it suffices, to show that the functions

$$f_i = \phi_i \circ (\phi \circ f \circ \theta^{-1}) : R^m \to R, \; i = 1, \ldots, n$$

are differentiable on an open subset D of $\theta(p)$. Thus if (U, θ) and (V, ϕ) are the corresponding coordinate systems, then we obtain the coordinate expression $\phi_i = f_i(\theta_1, \ldots, \theta_m)$ for $i = 1, \ldots, n$ which must be differentiable at $\theta(p) = (p_1, \ldots, p_m)$.

(3.4) *Definition*

f is said to be a C^m diffeomorphism of X onto Y if and only if f is a bijection of X onto Y, f is a C^m manifold morphism of X into Y and f^1 is a C^m manifold morphism of Y into X. The set of all C^m diffeomorphisms of X onto Y is denoted by Diff ^{m} (X). If a diffeomorphism of X onto Y exists, we say that X and Y are diffeomorphic.

A C^{∞} manifold morphism of smooth manifold is called a smooth morphism. The real line $\underline{R}$ with its standard differentiable structure determined by thc atlas consisting of the single chart $(\underline{R}, id_{\underline{R}})$ is a smooth manifold. We denote by $C^{\infty}(X)$ the set of all smooth, real valued functions on X and provide it with the obvious (pointwise) structure of a commutative algebra with identity.

(3.5) ***Examples***

(1) Let R be a manifold with the usual structure $\theta : R \to R : t \mapsto t$. Let (-1, 1) be an open submanifold of R then $f:(-1,1) \to R$ defined by $t \mapsto t/(t-t^2)$ is a diffeomorphism.

(2) Let R be the above manifold with the usual structure, and let M_1 be the manifold with coordinate function $\phi : M_1 \to R : t \mapsto t^3$. Then the map $f : M_1 \to R : s \mapsto s^3$ is a C^∞ homeomorphism and the inverse homeomorphism $f^{-1} : R \to M_1 : u \mapsto u^{1/3}$ is differentiable of class C^∞ relative to the above differentiable structure : for $t \in R$ we have the coordinate expression $(\phi \circ f^{-1} \circ \theta^{-1})(t) = (\phi \circ f^{-1})(t) = \phi(t^{1/3}) = t$, however note that identity map $g : M_1 \to R : t \mapsto t$ is not C^∞ since $(\theta \circ g \circ \phi^{-1})(t) = t^{1/3}$ which is not C^∞, that is, the identity map is not a diffeomorphism.

4

The Tangent Space

(4.1) *Tangent and Normal vectors*

We define individual tangent vector at a point on a surface, and then prove that this collection of vectors form a plane. Althogh we might think intuitively of tangent vectors at a point p on a surface as starting at the point p, for convenience we will consider all vectors as being translated so that they to start at the origin. The intuitive idea behind the following definition is the observation that any vector tangent to a smooth surface is tangent to some curve in the surface.

(4.2) *Definition*

Let $X \subset R^3$ be a smooth surface and let $p \in X$ be a point. A vector v in R^3 is a **tangent vector** to X at p if there is a curve $c:(-\varepsilon, \varepsilon) \to X$ for some number $\varepsilon > 0$ such that $c(0) = p$ and $c'(0) = v$. The collection of all tangent vectors to X at p is denoted T_pX, and it is called the **tangent plane** to X at p or tangent space to X at p.

For each tangent vector v as in the above definition there will be may corresponding curves c.

Examples:

(1) Consider R^2 as a smooth surface in R^3. Since the tangent vector to any curve in R^2 is itself in R^2, it is not hard to see that for each point $p \in R^2$ we have $T_pR^2 = R^2$.

(2) Let $p = \begin{pmatrix} 1 \\ 0 \\ 0 \end{pmatrix} \in S^2$, and let $v \in R^3$ be any vector of the form $v = \begin{pmatrix} 0 \\ a \\ b \end{pmatrix}$.

Letting $D = \sqrt{a^2 + b^2}$, then the curve $c:(-\pi, \pi) \to S^2$ given by

$$c(t) = \begin{pmatrix} \cos Dt \\ (a/D)\sin Dt \\ (b/D)\sin Dt \end{pmatrix}$$

has the property that $c(0) = p$ and $\dot{c}(0) = v$. We thus see that $T_p S^2$ contains the y-z plane. That $T_p S^2$ in fact equals the y-z plane will follow from lemma below.

The following lemma not only shows that the collection of tangent vectors at a point forms a plane, but it also gives an easy way of finding tangent planes.

(4.3) *Lemma*

Let $X \subset R^3$ be a smooth surface and let $p \in X$ be a point. Then $T_p X$ is a two-dimensional vector space. Moreover, if $\phi : U \to X$ is a coordinate patch such that $p \in \phi(U)$, then $T_p X$ is the subspace of R^3 spanned by the vectors $\{v_1, v_2\}$ evaluated at $\phi^{-1}(p)$.

Proof: It will suffice to prove the second part of the lemma. Let span $\{v_1, v_2\}$ denote the vector subspace of R^3 spanned by the vectors $\{v_1, v_2\}$, where v_1 and v_2 are evaluted at $\phi^{-1}(p)$; we will prove that $T_p X = \text{span}\{v_1, v_2\}$. First, suppose $v \in T_p X$, so that $v = \dot{c}(0)$ for some curve $c:(-\varepsilon, \varepsilon) \to X$. such that $c(0) = p$. We can assume without loss of generality that the image of c is entirely contained in $\phi(U)$. Let $c_1, c_2:(-\varepsilon, \varepsilon) \to R$ be the coordinate functions of the curve c with respect to ϕ. Then we see that

$$v = \dot{c}(0) = \dot{c}_1(0)\, v_1 + \dot{c}_2(0)\, v_2 \in \text{span}\{v_1, v_2\}.$$

Next, suppose $v \in \text{span}\{v_1, v_2\}$, so that $v = a\, v_1 + b\, v_2$ for some real number a and b without loss of generality we may assume that the cooordinate patch ϕ has been chosen so that $o_2 \in R^2$ is contained in the set U and that $\phi(o_2) = p$. Since U is open, all point in R^2 close enough to the origin are contained in U. We now define a curve $c : (-\varepsilon,\varepsilon) \to \phi(U) \subset X$ by the formula $c(t) = \phi\left(\begin{pmatrix} at \\ bt \end{pmatrix}\right)$ for some small enough number $\in$(to insure that the points $\begin{pmatrix} at \\ bt \end{pmatrix}$ are contained in U for all t $\in(-\varepsilon,\varepsilon)$). It is now straight forward to verify that $c(0) = p$ and $c`(0) = a\,v_1 + bv_2 = v$. Hence $v \in T_pX$.

(4.4) *Definition*

Let X be a differentiable manifold of dimension n. Let x be any point of X. A C^1 morphism c of an open interval $I \subset R$ into X with $0 \in I$ and $c(0) = x$ is called a C^1 path in X passing through x. Two C^1 path c_1 and c_2 in X passing through x are said to be tangent at x if and only if for some chart θ for X at x, $\theta \circ c_1$ and $\theta \circ c_2$ are tangent at 0, i.e., $D(\theta \circ c_1)(0) = D(\theta \circ c_2)(0)$, The tangency of paths at $x \in X$ is independent of the chart used (for by taking restrictions if necessary we may suppose $U_1 = U_2$. Hence we have $\theta_2 \circ c_1 = \left(\theta_2 \circ \theta_2^{-1}\right) \circ (\theta_2 \circ c_1)$. Thus by C^1 composite mapping theorem it follows that $D(\theta_2 \circ c_1)(0) = D(\theta_2 \circ c_2)(0)$. Thus we say c_1, c_2 are tangent at $x \in X$ with respect to θ, for any local chart at x. Thus it is evident that "C^1 path c_1 and c_2 in X passing through x are tangent at x" is an equivalence relation in the set of all C^1- paths in X passing through x. For each C^1 path c in X passing through x the equivalence class of c is denoted by $\underline{c}$. The set of all ordered pairs $(x, \underline{c})$, where $\underline{c}$ is a C^1 path in X passing through x, is denoted by $T_x(X)$ or $T_x X$ and known as tangent space at x in X. An element of $T_x X$ is called a tangent vector at x.

(4.4) Let X be a differentiable manifold of dimension n and $\underline{R}$ be a field of real numbers. Let x be any point of X and let θ be a chart for X at x. Then any two C^1 paths c_1 and c_2 in X passing through x are equivalent if and only if $D(\theta \circ c_1)(0) = D(\theta \circ c_2)(0)$. Hence we can define a map $\theta_x : T_x X \to \underline{R}^n$ by $\theta_x(x, \underline{c}) = D(\theta \circ c)(0)$, where c & $\underline{c}$ has their usual meaning defined above. We observe that θ_x is an injection of $T_x X$ into $\underline{R}^n$. To prove that θ_x is a surjection of $T_x X$ onto $\underline{R}^n$, Let ξ be any vector in $\underline{R}^n$, and Let c_ξ be the map $\underline{R} \ni a \mapsto \theta(x) + a\xi \in \underline{R}^n$ of $\underline{R}$ into $\underline{R}^n$. Put $c_{\xi\theta} = \theta^{-1} \circ \left(c_\xi \restriction c_\xi^{-1}[\mathrm{Im}(\theta)]\right)$. Then $c_{\xi\theta}$ is a C^1 morphism of the open neighbourhood $c_\xi^{-1}[Im(\theta)]$ of 0 in $\underline{R}$ into X and so a C^1 path in X passing through x such that $D(\theta \circ c_{\xi\theta})(0) = \xi$. Thus θ_x is a surjection of $T_x X$ onto $\underline{R}^n$.

(4.6) Let X be a differentiable manifold of dimension n. Let x be any point of X, and let θ and ψ be charts for X at x. Then for each ξ in $\underline{R}^n$

$$\psi_x\left(\theta_x^{-1}(\xi)\right) = \psi_x\left(x, \underline{c}_{\xi\theta}\right) = D\left(\psi \circ c_{\xi\theta}\right)(0) = D\left(\psi \circ \theta^{-1} \circ c_\xi\right)(0)$$

$$= D\left(\psi \circ \theta^{-1}\right)\left(c_\xi(0)\right).\, Dc_\xi(0) = D\left(\psi \circ \theta^{-1}\right)(\theta(x))(\xi)$$

Therefore $\psi_x \circ \theta_x^{-1} = D\left(\psi \circ \theta^{-1}\right)(\theta(x))$ and so $\psi_x \circ \theta_x^{-1}$ is a linear homeomorphism of $\underline{R}^n$ onto $\underline{R}^n$ itself. It follows that we can define a unique Banach space structure on $T_x X$ by means of θ_x such that θ_x^{-1} is a linear homeomorphism of $\underline{R}^n$ onto $T_x X$. For each pair of vectors v and w in $T_x X$ and each scalar a in $\underline{R}$ we define addition and scalar multiplication in $T_x X$ as follows

$$v + w = \theta_x^{-1}(\theta_x(v) + \theta_x(w)), \quad a\,v = \theta_x^{-1}(a\theta_x(v)), \; \|v\| = \|\theta_x(v)\|$$

Thus $T_x X$ becomes a Banach space endowed with the Banach space structure defined above. The dimensions of $T_x X$ as a vector space over $\underline{R}$ is the same as the dimension of X as a manifold.

Indeed, if (U,θ) is a chart at x in X with coordinate function $\theta^i(i=1,\ldots,n)$ then $\theta(x)=\left(\theta^1(x),\ldots,\theta^n(x)\right)$.

(4.7) Let X be an open subset of a Banach space S over $\underline{R}$. Consider X as an open submanifold of S with the only chart $\theta = id_X$. Then for each x in X the linear homeomorphism θ_x is called the canonical morphosim of $T_x X$ onto S, and is denoted by τ_x. $T_x X$ is then identified with S by means of τ_x.

5
Tangent Maps

(5.1) Let X and Y be differentiable manifolds modelled on a Banach space R and S over A respectively. Let f be a C^1 manifold morphism of X into Y. Let x be a point of X and (U,θ) be a chart for X at x and let (V,ϕ) be a chart for Y at $f(x)$ such that $f(U) \subset V$. Let c_1 and c_2 be any twc C^1 paths in X tangent at x; then $f \circ c_1$ and $f \circ c_2$ are c^1 paths in Y tangent at $f(x) \in Y$, for:

$$D(\phi \circ f \circ c_1)(0) = D\left(\phi \circ f \circ \theta^{-1} \circ \theta \circ c_1\right)(0)$$

$$= \left(D\left(\phi \circ f \circ \theta^{-1}\right)\left(\theta\left(c_1(0)\right)\right)\right)\left(D(\theta \circ c_1)(0)\right)$$

$$= \left(D\left(\phi \circ f \circ \theta^{-1}\right)\left(\theta\left(c_2(0)\right)\right)\right)\left(D(\theta \circ c_2)(0)\right)$$

$$\text{since } \left(D(\theta \circ c_1)(0)\right) = \left(D(\theta \circ c_2)(0)\right)$$

$$= D\left(\phi \circ f \circ \theta^{-1} \circ \theta \circ c_2\right)(0)$$

$$= D(\phi \circ f \circ c_2)(0)$$

Hence we define the tangent map of f at x, denote it by $Tx(f)$ or $T_x f : T_x X \to T_{f(x)} Y$, that assigns to each vector $(x, \underline{c})$ in $T_x X$, where c is a C^1 path in X passing through x, the value $\left(f(x), \underline{f \circ c}\right)$ in $T_{f(x)} Y$. We observe that :

$$(T_x f)(x,\underline{c}) = (f(x), \underline{f\circ c}) = \phi_{f(x)}^{-1}\circ\phi_{f(x)}(f(x), \underline{f\circ c})$$

$$= \phi_{f(x)}^{-1}\circ D(\phi\circ f\circ c)(0) = \phi_{f(x)}^{-1}\circ D(\phi\circ f\circ\theta^{-1}\circ\theta\circ c)(0)$$

$$= \phi_{f(x)}^{-1}\circ D(\phi\circ f\circ\theta^{-1})(\theta(x))\circ D(\theta\circ c)(0)$$

$$= \phi_{f(x)}^{-1}\circ D(\phi\circ f\circ\theta^{-1})(\theta(x))\circ\theta_x(x,\underline{c})$$

Hence :

$$T_x f = \phi_{f(x)}^{-1}\circ D(\phi\circ f\circ\theta^{-1})(\theta(x))\circ\theta_x \tag{5.1.1}$$

where $D(\phi\circ f\circ\theta^{-1})(\theta(x))$ is called the local expression of f at x relative to charts θ and ϕ. If X is an open subset of R and Y that of S then $T_x f = (\tau_{f(x)})^{-1}\circ Df(x)\circ\tau_x$.

(5.2) Let X be a differentiable manifold of dimension n. Let (U,θ) be a chart for X at a point $x\in X$. Then θ is a diffeomorphism of the open submanifold U of X onto the open submanifold $\theta(U)$ of $\underline{R}^n$ and so $T_x\theta$ is a linear homeomorphism of T_xU onto $T_{\theta(x)}\theta(U)$, which can be identified with $\underline{R}^n$. Moreover if we put $id_{\underline{R}^n} = \phi$, Then $T_x\theta = \phi_{\theta(x)}^{-1}\circ\theta_x$.

(5.3) Let f be a C^k manifold morphism of a C^k manifold X into a C^k manifold Y. Let g be another C^k manifold morphism of Y into a C^k manifold Z, then for each $x\in X$ one has

$$T_xX\xrightarrow{T_xf}T_{f(x)}Y\xrightarrow{T_{f(x)}g}T_{g(f(x))}Z,\quad T_xX\xrightarrow{T_x(g\circ f)}T_{g\circ f(x)}Z$$

and it can be proved that for all $x\in X$, $T_x(g\circ f) = T_{f(x)}g\circ T_xf$. If f is a diffeomorphism of X onto Y then for each $x\in X$, $(T_xf)^{-1} = T_{f(x)}f^{-1}$. It is evident that $T_x(id_X) = id_{T_xX}$ for all $x\in X$.

(5.4) Let X be a C^k manifold of dimensions n. Let f be a C^1 manifold morphism of X into a Banach Space S over A. For each $x\in X$ the continuous linear map $\tau_{f(x)}\circ T_x f$ of $T_x X$ into S is called the differential of f at x and is denoted by $d_x(f)$ or by $d_x f$ or even by $T_x f$. If X is an open subset of $\underline{R}$ then $d_x f$ will be identified with $Df(x)$. For a chart (U,θ) at x in X.

$$d_x f = D\left(f\circ\theta^{-1}\right)\left(\theta(x)\right)\circ\theta_x \tag{5.4.1}$$

In particular, $d_x\theta = \theta_x$, If ψ is another chart at x, then $d_x\psi = D\left(\psi\circ\theta^{-1}\right)\left(\theta(x)\right)\circ d_x\theta$.

If X is an open subset of R then $d_x f = Df(x)\circ\tau_x$ and so $d_x f$ can be identified with $Df(x)$. Let X be an open subset of $\underline{R}^n$, then for each $x\in Dm(\theta)$ and for each $i\in\underline{n}$, $d_x\theta^i = pr_i\circ\theta_x$, we denote $d_x\theta^i$ by dx^i by abuse of notation. Let $(e_i; i\in\underline{n})$ be the canonical basis of $\underline{R}^n$ then for each x in $Dm(\theta)$ and each pair of elements i and j in $\underline{n}$.

$$\left(dx^i\right)\left(\theta_x^{-1}\left(e_j\right)\right) = pr_i\circ\theta_x\left(\theta_x^{-1}\left(e_j\right)\right) = pr_i\left(e_j\right) = \delta_{ij} \tag{5.4.1(a)}$$

Hence $\left(dx^i; i\in\underline{n}\right)$ is a basis of the dual $(T_x X)^*$ of $T_x X$, dual to the basis $\left(\theta_x^{-1}(e_i); i\in\underline{n}\right)$ of $T_x X$. Consequently any vector v in $T_x X$ can be uniquely written as

$$v = \sum_{i\in\underline{n}}\left(d_x\theta^i\right)(v)\theta_x^{-1}(e_i) = \sum_{i\in\underline{n}}\left(dx^i\right)(v)\theta_x^{-1}(e_i) \tag{5.4.2}$$

Let α be another charts for X such that $Dm(\alpha) = Dm(\theta)$. Let $x\in Dm(\theta)$ and $j\in\underline{n}$, Then

$$\alpha_x^{-1}\left(e_j\right) = \sum_{i\in\underline{n}}\left(d_x\theta^i\right)\left(\alpha_x^{-1}\left(e_j\right)\right)\theta_x^{-1}(e_i) \tag{5.4.3}$$

we have for each $i\in\underline{n}$, $\left(d_x\theta^i\right)\left(\alpha_x^{-1}\left(e_j\right)\right) = pr_i\circ\theta_x\circ\alpha_x^{-1}\left(e_j\right)$

Now by using (5.4) and (4.3) we have

$$\begin{aligned}\left(d_x\theta^i\right)\left(\alpha_x^{-1}\left(e_j\right)\right) &= pr_i\left(\left(D\left(\theta\circ\alpha^{-1}\right)\left(\alpha(x)\right)\right)\left(e_j\right)\right)\\ &= pr_i\left(D_j\left(\theta\circ\alpha^{-1}\right)\left(\alpha(x)\right)\right)\end{aligned}$$

Hence,

$$\alpha_x^{-1}(e_j) = \sum_{i \in \underline{n}} pr_i\left(D_j\left(\theta \circ \alpha^{-1}\right)(\alpha(x))\right)\theta_x^{-1}(e_i) \qquad (5.4.4)$$

Since $(d_x \theta^i ; i \in \underline{n})$ is a basis of $(T_x X)^*$, there exists a unique list of sealars $(b_i ; i \in \underline{n})$ in R such that $d_x f = \sum_{i \in n} b_i d_x \theta^i$. Now for each

$$j \in \underline{n},\ (d_x f)\left(\theta_x^{-1}(e_j)\right) = \left(\sum_{i \in \underline{n}} b_i d_x \theta^i\right)\left(\theta_x^{-1}(e_j)\right) = b_j \text{ by (5.4.1(a))}$$

also by (5.4)

$$\begin{aligned}\left(d_x f\left(\theta_x^{-1} e_j\right)\right) &= \tau_{f(x)} \circ T_x f \circ \theta_x^{-1}(e_j) \\ &= \left(D\left(f \circ \theta^{-1}\right)(\theta(x)) \circ \theta_x\right)\left(\theta_x^{-1}(e_j)\right) \\ &= \left(D\left(f \circ \theta^{-1}\right)(\theta(x))\right)(e_j) \\ &= D_j\left(f \circ \theta^{-1}\right)(\theta(x))\end{aligned}$$

Hence

$$d_x f = \sum_{i \in \underline{n}} \left(D_i\left(f \circ \theta^{-1}\right)\right)(\theta(x))\, d_x \theta^i \qquad (5.4.5)$$

(5.5) Let f be a C^1 differentiable function of a differentiable manifold X of dimension n into a differentiable manifold Y of dimension p. Let $(e_j : j \in \underline{p})$ be the canonical basis of R^p. Let $x \in X$ and ϕ be a charts at $f(x)$ for Y. Then for each vector $v \in T_x X, (T_x f)(v)$ is in $T_{f(x)} Y$ and so

$$(T_x f)(v) = \sum_{j \in \underline{p}} \left(d_{f(x)} \phi^j\right)\left((T_x f)(v)\right)\phi_{f(x)}^{-1}(e_j) \text{ (by (5.4.2))}$$

$$= \sum_{j \in \underline{p}} \left((\tau)_{\phi^j(f(x))} \circ T_{f(x)} \phi^j \circ T_x f\right)(v)\phi_{f(x)}^{-1}(e_j) \text{ (by (5.4))}$$

$$= \sum_{j \in \underline{p}} (\tau)_{\phi^j(f(x))} \circ T_x(\phi^j \circ f)(v)\phi^{-1}_{f(x)}(e_j)$$

$$= \sum_{j \in \underline{p}} d_x(\phi^j \circ f)(v)\phi^{-1}_{f(x)}(e_j)$$

Let X be a smooth manifold of dimension n. Let $\overline{X}$ be an open subset of X. Let $C^\infty(X, \underline{R})$ be the commutative algebra over $\underline{R}$ of all smooth real valued functions on X into $\underline{R}$. Let x be any point of $\overline{X}$ for each vector v in $T_x\overline{X}$.

Define the map $L_{(x,v)}: C^\infty(\overline{X}, \underline{R}) \to \underline{R}$

by $L_{(x,v)}f = d_x f(v) \in \underline{R}$

If X is an open subset of $\underline{R}^n$ then for each f in $C^\infty(X, \underline{R})$ and each $v \in \underline{R}^n$ we identify $L_{(x,v)} f$ with $(Df(x))(v)$.

Let (U, θ) be a chart for X at x. Then for each C^k path c in X passing through x and each f in $C^\infty(\overline{X}, \underline{R})$

$$L_{(x,(x,\underline{c}))}(f) = d_x f(x, \underline{c})$$

$$= \left(D(f \circ \theta^{-1})(\theta(x))(\theta_x(x, \underline{c}))\right)$$

$$= \left(D(f \circ \theta^{-1})(\theta(c(0)))\right)\left(D(\theta \circ c)(0)\right)$$

$$= D(f \circ \theta^{-1} \circ \theta \circ c)(0)$$

$$= D(f \circ c)(0)$$

Therefore $L_{(x,(x,\underline{c}))}(f)$ is called the derivative of f along the tangent vector to c at x.

(5.6) If v is a non-zero vector in $T_x X$ then, $L_{(x,v)}$ is not zero and for all v in $T_x X$ and for f and g in $C^\infty(X, \underline{R})$, $L_{(x,v)}(fg) = L_{(x,v)}(f)g(x) + f(x)L_{(x,v)}(g)$. Thus $L_{(x,v)}$ is a derivation on $C^\infty(X, \underline{R})$. The image of $T_x X$ under the linear monomosphism $L_x: v \to L_{(x,v)}$ is identified with $T_x X$ and is referred to as the tangent space at x.

Let $(e_i; i \in \underline{n})$ be the canonical basis of $\underline{R}^n$. Then for each f in $C^\infty(X, \underline{R})$ and each $i \in \underline{n}$.

$$L_{(x, \theta_x^{-1}(e_i))}(f) = (d_x f)\left(\theta_x^{-1}(e_i)\right)$$
$$= \left(D\left(f \circ \theta^{-1}\right)(\theta(x))\right)(e_i)$$
$$= D_i\left(f \circ \theta^{-1}\right)(\theta(x))$$

which will be denoted by $\partial f(x)/\partial\theta^i$ or simply by $\partial f/\partial x^i$. For each $i \in \underline{n}, L_{(x, \theta_x^{-1}(e_i))}$ will be denoted by $\partial/\partial\theta^i$ or $\partial/\partial x^i$ and is an element of $T_x X$ which will be identified with $\theta_x^{-1}(e_i)$. Since $\left(\theta_x^{-1}(e_i); i \in \underline{n}\right)$ is a basis of $T_x X$, $\left(\partial/\partial x^i; i \in \underline{n}\right)$ is a basis of image of $T_x X$ under L_x.

(5.7) Let $f: X \to Y$ is a smooth map and $x \in X$, then f is said to be an immersion at x if $T_x f$ is one to one i.e. a monomorphism of $T_x X$ into $T_{f(x)} Y$ and an immersion if this is true at each $x \in X$. f is a submersion at $x \in X$ if $T_x f$ is onto, i.e., an epimorphism of $T_x X$ onto $T_{f(x)} Y$ and a submersion if this is true at each $x \in X$. An immersion that is also a homeomorphism onto its image is an imbedding. A point $y \in Y$ is a regular value of f if, for every $x \in f^{-1}(y)$, f is a submersion at x (this is the case, in particular, if $f^{-1}(y) = \phi$); otherwise, y is a critical value of f.

If X' is an open subspace of X and $\{(U_\alpha, \phi_\alpha)\}_{\alpha \in \wedge}$ is an atlas for X, then $\{(U_\alpha \cap X', \phi_\alpha | U_\alpha \cap X') : \alpha \in \wedge, U_\alpha \cap X' \neq \phi\}$ is an atlas for X' and, with the differentiable structure determined by this atlas, X' is called an open submanifold of X. Note that $\dim X' = \dim X$. More generally, if $\dim X = n$ and $1 \le k \le n$ is an integer, then a topological subspace X' of X is called a k-dimensional submanifold of X if, for each $x \in X'$, there exist a chart (U, ϕ) in the differentiable structure for X such that $\phi(U \cap X') = \left\{\left(x^1, \ldots, x^k, x^{k+1}, \ldots, x^n\right) \in \phi(U) : x^{k+1} = \ldots = x^n = 0\right\}$. For each such (U,ϕ), one obtains a chart $(U \cap X', \phi')$ for X', where ϕ' is $\phi \restriction U \cap X'$ followed by the projection of $\underline{R}^n = \underline{R}^k \times \underline{R}^{n-k}$ onto $\underline{R}^k$. The collection of all such $(U \cap X', \phi')$ is an atlas for X' and so determines a differentiable structure for X'. A o-dimensional sub-manifold of X is a discrete subspace of X. Restrictions of C^∞ maps on X to submanifolds are C^∞ with respect to this submanifold differentiable structure. According to inverse function theorem, if X and Y have the same dimension and $f: X \to Y$ is smooth, then $T_x f: T_x X \to T_{f(x)} Y$ is an isomorphism if and only if f is a local diffeomorphism at x, i.e., if and only if there exist open neighbourhoods V of x and W of $f(x)$ such that $f|V$ a diffeomorphism of V onto W.

If $f: X \to Y$ is an imbedding, then its image $f(x)$ is a submanifold of Y and, regarded as a map of X onto $f(X)$, f is a diffemorphism. On the otherhand, if $f: X \to Y$ is any smooth map and $y \in Y$ is a regular value of f (so that, in particular, $dim X \geq dim Y$) then $f^{-1}(y)$ is either empty or a submanifold of X of dimension $dim X - dim Y$.

Examples:

1. The map $\phi: \underline{R} \to \underline{R}^2$ given by $\phi(t) = (t, |t|)$ is not differentiable at $t = 0$.
2. The map $\phi: \underline{R} \to \underline{R}^2$ defined by $\phi(t) = (t^3, t^2)$ is differentiable but is not an immersion because $T_0 \phi: \underline{R} \to \underline{R}^2$ is the zero map that is not injective.
3. The map $\phi: (0, 2\pi) \to \underline{R}^2$ defined by $\phi(t) = (2cos(t - \pi/2), \sin 2(t - \pi/2))$ is an immersion but is not a embedding. The image $\phi((0.2\pi)) = X$ is an "eight". Also, the inclusion $i: X \to \underline{R}^2$ is not an embedding, So $X = \phi((0.2\pi))$ is not a submanifold of R^2.
4. The curve $\phi: (-3,0) \to \underline{R}^2$ given by

$$\phi(t) = \begin{cases} (0,-(t+2)) \text{ if } t \in (-3,-1) \\ \text{a regular curve for } t \in (-1,-1/\pi) \\ (-t,-\sin 1/t) \text{ if } t \in (-1/\pi,0) \end{cases}$$

is an immersion but is not an embedding. A neighbourhood of 0 = (0, 0) has infinitely many connected components if one considers the induced topology for the set $\phi(-3,0) \subset R^2$.

5. The map $\phi: \underline{R} \to \underline{R}^3$ defined by $\phi(t) = (\cos 2\pi t, \sin 2\pi t, t)$ is an embedding. The image $\phi(\underline{R})$ is homeomorphic to $\underline{R}$.
6. The image $\phi(\underline{R})$ of the map $\phi: \underline{R} \to \underline{R}^2$ give by $\phi(t) = (\cos 2\pi t, \sin 2\pi t)$ is $S^1 \subset \underline{R}^2$. The map ϕ is immersion but not an embedding since it is not injective. But $\phi(\underline{R}) = S^1$ a submanifolds of $\underline{R}^2$ if we consider the inclusion map $i: S^1 \to \underline{R}^2$.

(5.8) *Riemannian metric*

Suppose that at each point p of a manifold X a positive inner product $g_p(u, v)(u, v \in T_p(X))$ on $T_p(X)$ is given. For a local coordinate system $(x^1,\ldots,x^n)$ on U, we let $g_{ij}(q) = g_q\left((\partial/\partial x^i)_q, (\partial/\partial x^j)_q\right)$ for $q \in U$. Each $g_{ij}(q)(i, j = 1,\ldots,n)$ is a function on U, and the matrix $(g_{ij}(q))_{i,j=1,\ldots,n}$ is a positive definite symmetric matrix. If $(\overline{x}^1,\ldots,\overline{x}^n)$ is another coordinate system on U.

Let $\overline{g}_{ij}(q) = g_q\left((\partial/\partial\overline{x}^i)_q, (\partial/\partial\overline{x}^j)_q\right)$, then we have

$$\overline{g}_{ij}(q) = \sum_{k,l=1}^{n} \frac{\partial x^k}{\partial \overline{x}^i}(q)\frac{\partial x^l}{\partial \overline{x}^j}(q) g_{kl}(q)$$

$$g_{ij}(q) = \sum_{k,l=1}^{n} \frac{\partial \bar{x}^k}{\partial x^i}(q)\frac{\partial \bar{x}^l}{\partial x^j}(q)\bar{g}_{kl}(q)$$

Hence, if the functions g_{ij} on U are all of class C^r, then so are the functions $\bar{g}_{ij}$, and conversely. That is, the property that the g_{ij}'s are all of class C^r does not depend on the choice of the local coordinate system. If all the g_{ij}'s are of class C^r on a neighborhood of each point of X, then the map $g: p \to g_p$ which assigns a positive inner product g_p to $T_p(X)$ for each point p of X, is called a *Riemannian metric* of class C^r. g_p is called the value of g at p and the g_{ij}'s are called the components of g with respect to $(x^1,...,x^n)$. For simplicity, we shall restrict ourselves to *Riemannian metrics of class* C^∞, and call a

Riemannian metric of class C^∞ simply a Riemannian metric. When a Riemannian metric is given,we define the length $\|v\|$ of a tangent vector v at p by $\|v\|^2 = g_p(v, v)$.

If v has components $(\xi^1,...,\xi^n)$ with respect to $(x^1,...,x^n)$, then

$$\|v\|^2 = \sum_{i,j=1}^{n} g_{ij}(p)\xi^i\xi^j .$$

The Riemannian metric is often denoted by the symbol

$$ds^2 = \sum_{i,j} g_{ij}dx^i dx^j .$$

A manifold with a Riemannian metric is called a Reimannian manifold.

Let φ be a differentiable curve on a Riemannian manifold M defined on an open interval (a, b), and let v_t be the tangent vector of φ at $\varphi(t)$.Then $t \to \|v_t\|$ is a continuous function of t. If $a < c < d < b$, then we call $L(\varphi; c, d) = \int_c^d \|v_t\| dt$ the length of φ between $\varphi(c)$ and $\varphi(d)$. For a differentiable curve φ defined on a closed interval $[a, b]$, there is a differentiable curve φ' defined on an open interval (a', b') containing $[a, b]$ such that $\varphi(t) = \varphi'(t)$ for $t \in [a, b]$. We let $L(\varphi) = L(\varphi'; a, b)$, and call $L(\varphi)$ the length of φ. If φ is a piecewise differentiable curve, then it is a sum of a finite number of differentiable curves define on closed intervals, and its length is defined as the sum of the lengths of these differentiable curves.

6

Product Manifolds

(6.1) *Definition*

Let X and Y be smooth manifolds of dimension m and n, respectively. If (U,θ) is a chart on X and (V,ϕ) is a chart on Y, then $(U \times V,\ \theta \times \phi)$ is a chart on the topological space $X \times Y$. (The product map $\theta \times \phi : U \times V \to \underline{R}^m \times \underline{R}^n = R^{m+n}$ is defined by $(\theta \times \phi)(p,q) = (\theta(p), \phi(q))$. The set of all such charts is an atlas for $X \times Y$ determining the product manifold structure on $X \times Y$. The process obviously extends to larger (finite) products.

(6.2) The cartesian product of S^1 with itself is a two dimensional manifold called a torus $T^2 = S^1 \times S^1$ and can be thought of as the surface of an inner tube. The points on T^2 are given by pairs (θ, ρ) of angular coordinates, with two pairs being identified if they differ by integral multiple of 2π. In otherwords, (θ, ρ) and $(\tilde{\theta}, \tilde{\rho})$ describe the same points on T^2 if and only if : $\theta - \tilde{\theta} = 2k\pi$ and $\rho - \tilde{\rho} = 2l\pi$, for integers k, l. Thus T^2 can be covered by three coordinates charts

$$U_1 = \{(\theta, \rho) | 0 < \theta < 2\pi,\ 0 < \rho < 2\pi\},$$

$$U_2 = \{(\theta, \rho) | \pi < \theta < 3\pi,\ \pi < \rho < 3\pi\},$$

$U_3 = \left\{(\theta,\rho)\middle|\frac{\pi}{2} < \theta < \frac{5\pi}{2}, \frac{\pi}{2} < \rho < \frac{5\pi}{2}\right\}$.

The first overlap function is

$$\begin{aligned}\phi_1 \circ \phi_2^{-1}(\theta,\rho) &= (\theta,\rho),\ \pi < \theta < 2\pi,\ \pi < \rho < 2\pi \\ &= (\theta - 2\pi,\ \rho),\ 2\pi < \theta < 3\pi,\ \pi < \rho < 2\pi \\ &= (\theta,\ \rho - 2\pi),\ \pi < \theta < 2\pi,\ 2\pi < \rho < 3\pi \\ &= (\theta - 2\pi,\ \rho - 2\pi),\ 2\pi < \theta < 3\pi,\ 2\pi < \rho < 3\pi\end{aligned}$$

On the intersection $(U_1 \cap U_2)$ which is the set of all (θ,ρ) with neither θ nor ρ being an integral multiple of π. More generally an m-dimensional torus is given by the m-fold cartesian product $T^m = S^1 \times ... \times S^1$ of S^1 with itself.

(6.3) *Example:*

Show that if X is m-dimensional manifold and Y is n -dimensional manifold, then the Cartesian product X × Y has the structure of an (m + n) dimensional manifold.

Solution:

Suppose X has coordinate chart $\theta_i : U_i \to X$ and Y has coordinate chart

$\Psi_j : V_j \to Y$ with U_i open in R^m and V_j open in R^n. We form coordinate chart on X × Y by taking

$\theta_i \times \Psi_j : U_i \times V_j \to X \times Y$, defined by $(u, v) \to (\theta_i(u), \Psi_j(v))$.

Condition 2 of (2.8) is easily verified as follows : for every $(x, y) \in X \times Y$, we know that for some i, $x \in \theta_i(U_i)$ and for some j, $y \in \Psi_j(V_j)$, so $(x, y) \in (\theta_i \times \Psi_j)(U_i \times V_j)$.

To verify condition 3, take two coordinate patches.

$\theta_i \times \Psi_j : U_i \times V_j \to X \times Y$ and $\theta_k \times \Psi_l : U_k \times V_l \to X \times Y$

which overlap on

$W = (\theta_i \times \Psi_j)(U_i \times V_j) \cap (\theta_k \times \Psi_l)(U_k \times V_l)$

we must verify that $(\theta_i \times \Psi_j)^{-1}$ and $(\theta_k \times \Psi_l)^{-1}$ (W) are open, and that

$$F = (\theta_k \times \Psi_l)^{-1} \circ (\theta_i \times \Psi_j) : (\theta_i \times \Psi_j)^{-1}(W) \to (\theta_k \times \Psi_l)^{-1}(W)$$

is infinitely differentiable. To do this, the reader must satisfy him or herself that, given invertible functions f, g, h and t with appropriate range and domains so that the compositions $f \circ h$ and $g \circ t$ make sense, the following equations hold :

$$(f \times g)^{-1} = f^{-1} \times g^{-1}, (f \times g) \circ (h \times t) = (f \circ h) \times (g \circ t)$$

Then we can see that $(\theta_i \times \Psi_j)^{-1}(W) = \theta_i^{-1}(W) \times \Psi_j^{-1}(W)$ is the Cartesian product of two open sets, which is open, similarly, $(\theta_k \times \Psi_l)^{-1}(W)$ is open. As for the transition function

$$F = (\theta_k \times \Psi_l)^{-1} \circ (\theta_i \times \Psi_j) = (\theta_k^{-1} \circ \theta_i) \times (\Psi_l^{-1} \circ \Psi_j)$$

which is the Cartesian product of two infinitely differentiable functions, making F infinitely differentiable.

(6.4) Let X_1 and X_2 are $C^k (k \in \underline{N})$ manifolds and let (x_1, x_2) be any point of the product manifold $X_1 \times X_2$. Then pr_i is a C^k submersion of $X_1 \times X_2$ onto X_i for $i = 1,2$.

(6.5) The function $v \mapsto \left(\left(T_{(x_1,x_2)}(pr_1) \right)(v), \left(T_{(x_1,x_2)}(pr_2) \right)(v) \right)$ is a linear homeomorphism of $T_{(x_1,x_2)}(X_1 \times X_2)$ onto $T_{x_1}(X_1) \times T_{x_2}(X_2)$, called canonical, and the two Banach spaces are often be identify by means of the above linear homeomorphism.

(6.6) Let f be a C^k morphism of $X_1 \times X_2$ into a C^k manifold Y. Then for each point (x_1, x_2) of $X_1 \times X_2$ and each point (v_1, v_2) of $T_{x_1}(X_1) \times T_{x_2}(X_2)$, which we canonically identify with $T_{(x_1,x_2)}(X_1 \times X_2)$,

$$\begin{aligned}\left(T_{(x_1,x_2)} f\right)(v_1, v_2) &= \left(T_{x_1}(f_{\cdot x_2})\right)\left(T_{(x_1,x_2)}(pr_1)\right)(v_1, v_2) + \\ &\qquad \left(T_{x_2}(f_{x_1})\right)\left(T_{(x_1,x_2)}(pr_2)\right)(v_1, v_2) \\ &= \left(T_{x_1}(f_{x_2\cdot})\right)(v_1) + \left(T_{x_2}(f_{x_1})\right)(v_2)\end{aligned}$$

(6.7) For $i = 1,2$, let f_i be a map on a C^k manifold Y into X_i, then the function $f :, y \mapsto (f_1(y), f_2(y))$ is a C^k morphism of Y into $X_1 \times X_2$ iff f_1 and f_2 are C^k morphism and for each y in Y and each v in $T_y Y$,

$$(T_y f)(v) = ((T_y f_1)(v), (T_y f_2)(v))$$

(6.8) For $i = 1,2$, let g_i be a C^k morphism of the manifold X_i into a C^k manifold Y_i, and let g be the map $\mathrm{X}_1 \times \mathrm{X}_2 \ni (\mathrm{x}_1, \mathrm{x}_2) \mapsto (\mathrm{g}_1(\mathrm{x}), \mathrm{g}_2(\mathrm{x})) \in \mathrm{Y}_1 \times \mathrm{Y}_2$, then g is a C^k morphism of $X_1 \times X_2$ into $Y_1 \times Y_2$.

(6.9) *Theorem*

Let f be a C^k morphism of a C^k manifold X into C^k manifold Y. Then the graph $\mathrm{Gr}(f)$ of f is a closed submanifold of the product manifold $X \times Y$ and the map $g : x \mapsto (x, f(x))$ is a C^k diffeomorphism of X onto $\mathrm{Gr}(f)$.

Proof: By topology g is a homeomorphism of X onto the subspace $\mathrm{Gr}(f)$ of $X \times Y$. Let c be the canonical isomorphism of $T_{(x, f(x))} X \times Y$ onto $(T_x X) \times (T_{f(x)} Y)$, x being an arbitrary point of X, then for each v in $T_x X$,

$$\mathrm{c}((T_x g)(v)) = ((T_x (id_X))(v), (T_x f)(v)) = (v, T_x f(v))$$

and so $T_x g$ is a monomorphism of $T_x X$ into $T_{g(x)}(X \times Y)$, and its image is a topological direct summand of $T_{g(x)}(X \times Y)$. Hence $\mathrm{Gr}(f)$ is a submanifold of $X \times Y$ and g is a C^k diffeomorphism of X onto $\mathrm{Gr}(f)$.

7

Vector Bundles

(7.1) Let X and Y be C^k ($k \in \underline{N}$) manifolds and η be a C^k surjective manifold morphism of X onto Y. Then the ordered triple (η, Y, X) is called a C^k bundle, Y is called the bundle space, X is called the base space, and η is called the bundle map of the C^k bundle (η, Y, X). We also define the C^k bundle by the bundle map η. Let (η, Y, X) be a C^k bundle, then (η, Y) is called a C^k bundle over X. For each $x \in X, \eta^{-1}[x] = \{y \in Y \mid \eta(y) = x\}$ is called the fibre of η over x, and is often denoted by Y_x.

For each $y \in Y$, $T_y\left(Y_{\eta(y)}\right)$ is identified with the vector subspace $\ker\left(T_y \eta\right)$ of $T_y Y$ and an element of $T_y\left(Y_{\eta(y)}\right)$ is called a vertical tangent vector at y.

Let (η, Y, X) and (η', Y', X') be two C^k bundles and let f be a C^k morphism of X into X' and g be C^k morphism of Y into Y' such that $f \circ \eta = \eta' \circ g$, i.e., the diagram give below commutes,

$$\begin{array}{ccc} Y & \xrightarrow{\eta} & X \\ {\scriptstyle g}\downarrow & & \downarrow{\scriptstyle f} \\ Y' & \xrightarrow{\eta'} & X' \end{array}$$

Fig. 7.1

Then the ordered pair (f, g) is called a morphism of the bundle η into the bundle η'. The morphism (f, g) is called an isomorphism of η onto η' if f and g are C^k diffeomorphisms. Let (η, Y, X) and (η', Y', X) be two C^k bundles over the same base space X and let g be a C^k morphism of Y into Y' such that $\eta' \circ g = \eta$. Then (id_X, g) is called a morphism of the bundle η into the bundle η' (relative to X).

Let X and S be C^k $(k \in \underline{N})$ manifolds. Let pr_1 be the projection of the product manifold $X \times S$ onto X. Then $(pr_1, X \times S, X)$ is a C^k bundle, called C^k trivial bundle over X of fibre type S. A C^k bundle (η, Y, X) is said to be trivializable iff there is an isomorphism of (η, Y, X) onto a C^k trivial bundle over X relative to X.

(7.2) Let (η, Y, X) be a C^k $(k \in \underline{N})$ bundle and let S be a C^k manifold, then an ordered pair (W, θ) of an open subset W of X and C^k diffeomorphsm θ of an open submanifold $\eta^{-1}[W]$ onto the product manifold $W \times S$ is called a trivializer for the bundle η of fibre type S iff $\eta | \eta^{-1}[W] = pr_1 \circ \theta$, i.e.,

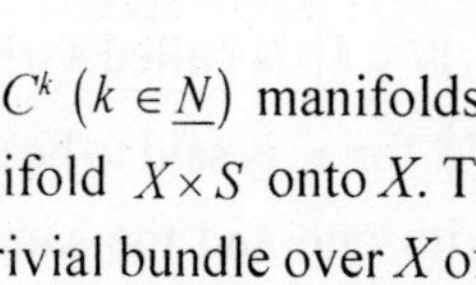

Fig. 7.2

The above diagram is commutative.

Let (W, θ) be a trivilizer for the bundle η of fibre type S. Then θ becomes an isomorphism of the bundle $(\eta \upharpoonright \eta^{-1}[W], \eta^{-1}[W], W)$ onto the tivial bundle $(pr_1, W \times S, X)$ relative to W. For each $x \in W$, we define a function ${}_x\theta : Y_x \to S$, which assigns to each $y \in Y_x$, the point $pr_2 \circ \theta(y) \in S$, where pr_2 is the projection of $W \times S$ onto S. For each $x \in W$; ${}_x\theta$ is a C^k diffeomorphism of the submanifold Y_x of Y onto the manifold S, moreover for each $y \in Y$, $\theta(y) = (\eta(y), {}_x\theta(y))$.

Let (W, θ) and (W', θ') be two trivilializers for η of fiber type

S and S' such that $W \cap W' \neq \phi$. Then for each points $(x, s) \in (W \cap W') \times S$.

$\theta'(\theta^{-1}(x, s)) = (x, {}_x\theta'({}_x\theta)^{-1}(s))$ and so ${}_x\theta' \circ {}_x\theta^{-1}$ is a C^k diffeomor-phism of S onto S'. Let $\{\alpha_i : i \in I\}$ be a C^k atlas for X. For each $i \in I$, put $W_i = Dm(\alpha_i)$ and let θ_i be a C^k diffeomorphism of $\eta^{-1}[W_i]$ onto $W_i \times S_i$ such that (W_i, θ_i) is a trivializer for η of fiber type S_i. Then $\{(W_i, \theta_i) : i \in I\}$ is called a trivialization for η. A trivialization $\{(W_i, \theta_i) : i \in I\}$ for η is said to be of fiber type S iff for all $i \in I$, S_i is identical with one and the same manifold S. A C^k bundle is said to be locally trivial iff it has a trivialization.

Let X and S be C^k manifolds. Let $\{\alpha_i : i \in I\}$ be a C^k atlas for X. Then $\{(Dm(\alpha_i), \theta_i); i \in I\}$ where θ_i is the identity map of $Dm(\alpha_i) \times S$, is a trivialization for the C^k trivial bundle $(pr_1, X \times S, X)$ of fiber type S, called canonical.

(7.3) A C^k $(k \in \underline{N})$ bundle (η, Y, X) is called a vector bundle iff for each $x \in X$, Y_x, the fiber over x, is a vector space over A.

Let (η, Y, X) be a C^k vector bundle. Then a trivialization $\{(W_i, \theta_i); i \in I\}$ for the bundle η of fiber type S_i, where S_i is a Banach Space over A for each $i \in I$, is called a trivialization for the vector bundle η of fiber type $\{S_i : i \in I\}$ iff for each $i \in I$ and each $x \in W_i$, Y_x is a Banach space over A, ${}_x\theta_i$ is a linear homeomorphism of Y_x onto S_i, and for each pair of elements i and j in I such that $W_i \cap W_j \neq \phi$ the function

$$W_i \cap W_j \ni x \mapsto ({}_x\theta_j) \circ ({}_x\theta_i)^{-1} \in L(S_i, S_j)$$

is a C^k morphism of the open submanifold $W_i \cap W_j$ of X into the Banach space $L(S_i, S_j)$ of all continuous linear maps of S_i into S_j with the strong topology.

Let (η, Y, X) be a C^k vector bundle with trivialization $\{(W_i, \theta_i); i \in I\}$ of fiber type S where S is a Banach space over A and (η', Y', X') be another C^k vector bundle with trivialization $\{(W'_j, \theta'_j; j \in I)\}$ of fiber type S where S is a Banach Space over A. Let (f, g) be a C^k morphism of the bundle η into the bundle η'. Then (f, g) is called a vector bundle morphism iff for each $x \in X$, $g|Y_x$ is a continuous linear map of the Banach space Y_x into Banach space $Y'_{f(x)}$, and there is a trivializer (W_i, θ_i) for some $i \in I$ and a trivializer (W'_j, θ'_j) for some $j \in I'$ such that $x \in W_i, f(x) \in W'_j, f[W_i] \subset W'_j$, and the function $x \mapsto ({}_x\theta'_j) \circ (g|Y_x) \circ ({}_x\theta_i)^{-1}$ is a morphism of W_i into $L(S, S')$. Now suppose that $X = X'$ and $f = id_X$. Then g is called a vector bundle morphism relative to X.

Let (η, Y, X) and (η', Y', X') be two C^k vector bundles. The set of all vector bundle morphisms of η into η' relative to X is denoted by *Hom*(η, η'), where *Hom* (η, η') is a vector space over A.

A vector bundle morphism of a C^k trivial vector bundle $(pr_1, X \times S, X)$ into a C^k trivial vector bundle $(pr_1, X \times S, X)$ is of the form $(x, s) \mapsto (f(x), g(x)(s))$ Where f is a C^k morphism of X into X, and for each x in X, $g(x)$ is a continuous linear map of S into S and the function $X \ni x \mapsto g(x) \in L(S, S')$ is C^k.

(7.4) Let X be a C^k $(k \in \underline{N})$ manifold modelled on $A^n (n \in \underline{N})$ and let (η, Y, X) be a C^k bundle with trivialization $\{(W_i, \theta_i); i \in I\}$ of fiber type $A^p (p \in \underline{N})$. If (η, Y, X) is a vector bundle then the following statements are equivalent.

(i) $\{(W_i, \theta_i); i \in I\}$ is a trivialization for the vector bundle (η, Y, X) of fiber type A^p.

(ii) For each $i \in I$ and $x \in W_i$, ${}_x\theta_i$ is a linear isomorphism of Y_x onto A^p.

(iii) For each $i \in I$ there exists a list $(S_j; j \in \underline{p})$ of C^k function on W_i into Y such that for each $j \in \underline{p}$. $\eta \circ S_j = id_{W_i}$ and the function $(x, y) \mapsto \sum_{j \in \underline{p}} pr_j(y) S_j(x)$ is a diffeomorphism of $W_i \times A^p$ onto $\eta^{-1}[W_i]$, where pr_j in the jth projection of A^p onto A.

(7.5) Let X and S be a C^k $(k \in \underline{N})$ manifolds. For each $i \in I$, Let W_i be any open subset of X such that $X = \bigcup_{i \in I} W_i$ and for each elements $(i, j) \in I \times I$, $W_i \cap W_j \neq \phi$. Let g_{ij} be a function on $W_i \cap W_j$ into a group G of C^k diffeomorphism of S onto itself, such that for each $x \in W_i \cap W_j \cap W_k$, $g_{ik}(\mathbf{x}) = g_{jk}(x) \circ g_{ij}(x)$. Then $(W_i, g_{ij})_{i,j \in I}$ is called a system of transition function for X and S with structure group G.

Let S be a Banach space over A. Then a system of transition function $(W_i, g_{ij})_{i,j \in I}$ is said to be a system of transition function for X of vector space type S iff for each elements $(i, j) \in I \times I$, $W_i \cap W_j \neq \phi$ and for each $x \in W_i \cap W_j$, $g_{ij}(x) \in G$, where G is a group of linear homeomorphism of S onto itself, and g_{ij} is a C^k function on $W_i \cap W_j$ into the Banach space $L(S, S)$.

Let $(W_i, g_{ij})_{i,j \in I}$ be a system of transition function for $C^k (k \in \underline{N})$ manifolds X and S. Define

$Y' = \bigcup_{i \in I} W_i \times S \times \{i\}$ and endow it with the sum topology.

For each element $(i, j) \in I \times I$, Let h_{ij} be the function

$$(W_i \cap W_j) \times S \ni (x, f) \mapsto (x, (g_{ij}(x))(f)) \in (W_i \cap W_j) \times S$$

then h_{ij} is a C^k diffeomorphism of $(W_i \cap W_j) \times S$ onto itself.

Define a relation R in Y' such that for each pair of elements (x, f, i) and (x', f', j) in Y', $(x, f, i) R (x', f', j) \Leftrightarrow x = x'$ and $(g_{ij}(x))(f) = f'$. Then R is an equivalence relation in Y'. Put $Y = Y'/R$. Endow Y with the quotient topology and let q be the quotient map of Y' onto the quotient space Y'/R. For each $i \in I$ the restriction $q_i = q \upharpoonright W_i \times S \times \{i\}$ is a homeomorphism of $W_i \times S \times \{i\}$ onto the open subset $q[W_i \times S \times \{i\}]$ of Y and so we transport the atlas for

$W_i \times S \times \{i\}$ to $q[W_i \times S \times \{i\}]$ by means of $q \restriction W_i \times S \times \{i\}$ and then an atlas for Y is defined.

Finally we define a function η on Y onto X in the following manner. For each $y \in Y$ there is an $i \in I$ and there is an element $(x, f, i) \in W_i \times S \times \{i\}$ such that $y = q(x, f, i)$,and we set $\eta(y) = x$, x being independent of i. Thus (η, Y, X) is a C^k bundle with trivialization $\{(W_i, q_i^{-1}); i \in I\}$. If S is a Banach space over A and if $(W_i, g_{ij})_{i,j \in I}$ is a transition system for X of vector space type S, then (η, Y, X) is a C^k vector bundle with trivialization $\{(W_i, q_i^{-1}); i \in I\}$.On the other hand let (η, Y, X) be a C^k bundle with trivialization $\{(W_i, \theta_i); i \in I\}$ of fiber type S.

Let I' be the set of ordered pairs $(i, j) \in I \times I$, $W_i \cap W_j \neq \phi$. For each member $(i, j) \in I \times I$; Let g_{ij} be the function on $W_i \cap W_j$ which assigns to each $x \in W_i \cap W_j$ the diffeomorphism $({}_x\theta_j) \circ ({}_x\theta_i^{-1})$ of S onto itself. Then $(W_i, g_{ij})_{i,j \in I}$ is a system of transition functions for X and S, and is said to be induced by the trivialization $\{(W_i, \theta_i); i \in I\}$ for (η, Y, X), the structure group being the group of all C^k diffeomorphism of S onto itself.

(7.6) Let k be any natural number and let S be a Banach space over A. Let X be a C^k manifold with atlas $A = \{(U_i, a_i); i \in I\}$ modelled on a Banach space R over A. Let Y be any set and let η be a surjection of Y onto X. For each $i \in I$, let θ_i be a bijection of $\eta^{-1}[U_i]$ onto $U_i \times S$. Such that $\eta \restriction \eta^{-1}[U_i] = pr_i \circ \theta_i$, where pr_i is the projectin of $U_i \times S$ onto U_i, then $(\alpha_i \times id_s) \circ \theta_i$ is a bijection of $\eta^{-1}[U_i]$ onto the open subset $\alpha_i[U_i] \times S$ of the product space $R \times S$ and so a chart for Y modelled on $R \times S$. Hence $B = \{(\alpha_i \times id_s) \circ \theta_i; i \in I\}$ is a C^k atlas for Y modelled on $R \times S$ iff for each pair of elements i and j in I the transition map induced by the pair of charts $(\alpha_i \times id_s) \circ \theta_i$ and $(\alpha_j \times id_s) \circ \theta_j$ is in $Diff^k(R \times S)$.

$\{(\alpha_i, \theta_i); i \in I\}$ is called a framing for η iff B is a C^k atlas of Y modelled on $R \times S$. A member of framing for η is called a frame for η.

Suppose that $\{(\alpha_i, \theta_i); i \in I\}$ is a framing for η. Then (η, Y, X) is a C^k bundle, Y with the C^k atlas B is the bundle space, X with A is the base space and η is the C^k bundle map of the bundle manifold Y onto the base manifold X, $\{(U_i, \theta_i); i \in I\}$ is a trivilization for the C^k bundle η of fibre type S.

Every local trivial bundle has *a* framing. Now suppose that the framing $\{(\alpha_i, \theta_i); i \in I\}$ satisfies the condition that for each element $(i, j) \in I \times I$ such that $U_i \cap U_j \neq \phi$ and for each $x \in U_i \cap U_j$ the function

$$q_{ij}(x): s \mapsto pr_2\left(\theta_j\left(\theta_i^{-1}(x, s)\right)\right).$$

where pr_2 is the projection of $U_j \times S$ onto S, is a linear homeomorphism of S onto itself, and the function $x \mapsto q_{ij}(x)$ is a C^k morphism of the manifold $U_i \cap U_j$ into the manifold $L(S, S)$. Then (η, Y, X) is a C^k vector bundle with framing $\{(\alpha_i, \theta_i); i \in I\}$ and trivilization $\{(U_i, \theta_i); i \in I\}$ of fibre type S.

Every C^k vector bundle has a framing.

(7.7) Let R and S be any two Banach space over A. Let R^* be the strong dual of R. Let $T_j^i(R, S)$ be the set of all continuous $(i + j)$ multilinear maps of the Banach space $\left(R^*\right)^i \times R^j$ into S, i & j being any natural numbers. Then $T_j^i(R, S)$ is a Banach space with the norm defined in the usual way. For an element f of $T_j^i(R, S)$ we define

$$\| f \| = sup\left(\| f\left(r_1^*, r_2^*, \ldots, r_i^*, r_1, \ldots, r_j\right) \|\right)$$

An element of $T_j^i(R, S)$ is called a tensor on R with values in S, Contravariant of order i and covariant of order j or simply of type (i, j). We write $T_j^i(R) = T_j^i(R, A)$, we denote by $T_o^i(R, S)$ the set of all continuous i-multilinear maps of $\left(R^*\right)^i$ into S, by $T_j^o(R, S)$ the set of all continuous *j*- multilinear maps of R^j into *S*, and by $T_o^o(R,S)$ the Banach space $clm(A,S)$. We write $T_o^i(R) = T_o^i(R, A)$ and $T_j^o(R) = T_j^o(R, A)$. An element of $T_o^i(R, S)$ is called a contravariant tensor of order *i* with values in *S* and an element of $T_j^o(R,S)$ is called a covariant tensor of order *j* with values in *S*. We denote $T_1(R,S)$ by $L(R,S)$. Evidently $T_1(R) = L(R, A) = R^*$.

Let *g* be a linear homeomorphism of *R* onto a Banach space *R*. We define a map $T_j^i(g)$ of $T_j^i(R,S)$ into $T_j^i(R',S)$ as follows, for each *f* in $T_j^i(R,S)$ and each point $(r_1^*,\ldots,r_i^*)$ in $(R'^*)^i$ and each points $(r_1,\ldots,r_j)$ in R'^j

$$\left(\left(T_j^i(g)\right)(f)\right)\left(r_1^*,\ldots,r_i^*,r_1,\ldots,r_j\right)$$

$$= f\left(r_1^* \circ g,\ldots,r_i^* \circ g, g^{-1}(r_1),\ldots,g^{-1}(r_j)\right)$$

then $T_j^i(g)$ is a linear homeomorphism of $T_j^i(R,S)$ onto $T_j^i(R',S)$.

Let *X* be a C^k $(k \in N)$ manifold modelled on a Banach space *R* over *A* and let (η, Y, X) be a C^∞ vector bundle with framing $\{(\theta_m, \psi_m); m \in I\}$ of fibre type *R*. For each subset *W* of *X* put $T_j^i\left(\eta^{-1}[W],S\right) = \bigcup_{x \in W} T_j^i\left(\eta^{-1}(x),S\right)$.

Let $T_j^i(\eta)$ be the function on $T_j^i(Y,S) = T_j^i\left(\eta^{-1}(X),S\right)$ onto *X* such that for each *x* in *X* and each *f* in $T_j^i\left(\eta^{-1}(x),S\right), \left(T_j^i(\eta)\right)(f) = x$. For each *m* in *I* put $U_m = Dm(\theta_m)$. For each *m* in *I* and each *x* in U_m, ${}_x\psi_m : y \mapsto pr_2\left(\psi_m(y)\right)$ is a linear homeomorphism of $\eta^{-1}(x)$ onto *R* and so $T_j^i({}_x\psi_m)$ is a linear homeomorphism of $T_j^i\left(\eta^{-1}(x),S\right)$ onto $T_j^i(R,S)$. For each *m* in *I*. Let $T_j^i(\psi_m)$ be the bijection of $T_j^i\left(\eta^{-1}[U_m],S\right)$ onto $U_m \times T_j^i(R,S)$ such that for each *x* in U_m and each *f* in $T_j^i\left(\eta^{-1}(x),S\right)$, $\left(T_j^i(\psi_m)\right)(f) = \left(x, \left(T_j^i({}_x\psi_m)\right)(f)\right)$. Then for each *m* in $I\left(T_j^i(\eta)\right)^{-1}[U_m] = T_j^i\left(\eta^{-1}[U_m],S\right)$ and so $T_j^i(\eta)\Big|\left(T_j^i(\eta)\right)^{-1}[U_m] = pr_1 \circ T_j^i(\psi_m)$, where pr_1 is the projection of

$U_m \times T_j^i(R,S)$ onto $U_{m'}$, for each pair of elements m and m' in I, each x in $U_m \cap U_{m'}$ and each $h \in T_j^i(R,S)$, $\left(T_j^i(\psi_{m'})\right)\left(\left(T_j^i(\psi_m)\right)\right)^{-1}(x,h) = \left(x, \left(T_j^i\left({}_x\psi_{m'} \circ {}_x\psi_m^{-1}\right)\right)(h)\right)$, so that $(x,h) \mapsto \left(T_j^i(\psi_{m'})\right)\left(\left(T_j^i(\psi_m)\right)^{-1}(x,h)\right)$ is a C^∞ diffeomorphism of $(U_m \cap U_{m'}) \times T_j^i(R,S)$ onto itself and for each x in $U_m \cap U_{m'}$. $h \mapsto pr_2\left(\left(T_j^i(\psi_{m'})\right)\left(\left(T_j^i(\psi_m)\right)^{-1}(x,h)\right)\right)$ is a linear homeo-morphism of $T_j^i(R,S)$ onto itself, and the function $x \mapsto \left(h \mapsto pr_2\left(\left(T_j^i(\psi_{m'})\right)\left(T_j^i(\psi_m)\right)^{-1}(x,h)\right)\right)$ is a C^∞ morphism of $U_m \cap U_{m'}$ into $L\left(T_j^i(R,S), T_j^i(R,S)\right)$. Hence by (7.6) ($T_j^i(h)$, $T_j^i(Y, S)$, X) is a C^∞ vector bundle with framing $\left\{\left(\theta_m, T_j^i(\psi_m)\right); m \in I\right\}$ of fibre type $T_j^i(R,S)$.

Let $AT_j(R,S)$ be the Banach space over A of all alternating continuous j- multilinear maps of R^j into S with the norm defined in the usual way. We define $AT_j(Y,S)$ and $AT_j(\eta)$ similarly. Then $\left(AT_j(\eta), AT_j(Y,S), X\right)$ is a C^∞ vector bundle with framing $\left\{\left(\theta_m, AT_j(\psi_m)\right); m \in I\right\}$ of fibre type $AT_j(R,S)$. If $S = A$, then we write

$$AT_j(R,S) = AT_j(R),\ T_j^i(Y,S) = T_j^i(Y) \text{ and}$$
$$AT_j(Y,S) = AT_j(Y).$$

8

Tangent Bundle

(8.1) Let k be any positive integer greater than 1 or equal to ∞. Let X be a C^k manifold with an atlas α modelled on a Banach space R over A, put $TX = \bigcup_{x \in X} T_x X$. Let τ_X be the function on TX onto X, called the foot map such that for each x in X and each v in $T_x X$ $\tau_X(v) = x$. For each Chart (U, θ) in α Let ${}_\theta\tau$ be the function on $\tau_X^{-1}[U]$ onto $U \times R$ such that for each x in U and each C^k path c passing through x in X ${}_\theta\tau(x, \underline{c}) = (x, D(\theta \circ c)(0))$ then $\tau_X | \tau_X^{-1}[U] = pr_1 \circ {}_\theta\tau$

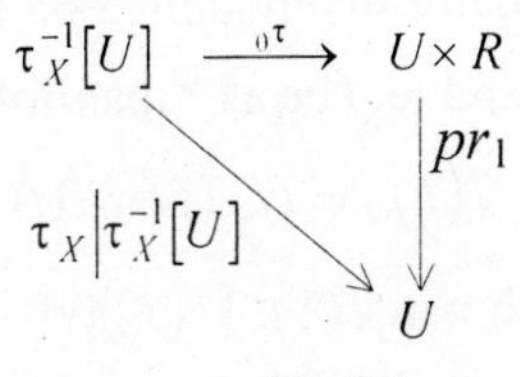

Fig. 8.1

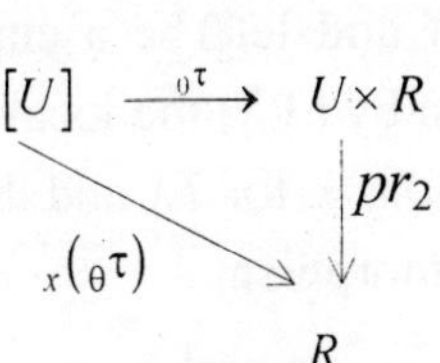

Fig. 8.2

where pr_1 is the projection of $U \times R$ onto U, ${}_{\theta}\tau$ is a bijection of $\tau_X^{-1}[U]$ onto $U \times R$, and for each x in U and each v in $T_x X$, ${}_{\theta}\tau(v) = (x, \theta_x(v))$, so that ${}_x({}_{\theta}\tau) = \theta_x$ interms of the notation of (4.4) and (7.2).

Let (U,θ) and (V,β) be any two charts in atlas $\wedge$ such that $U \cap U \neq \phi$. Then by (4.6) for each point (x, r) in $(U \cap V) \times R$,

$$ {}_{\beta}\tau \circ {}_{\theta}\tau^{-1}(x,r) = \left(x, D\left(\beta \circ \theta^{-1}\right)(\theta(x))(r)\right) $$

hence ${}_{\beta}\tau \circ {}_{\theta}\tau^{-1}$ is a diffeomorphism of $(U \cap V) \times R$ onto itself moreover for each x in $U \cap V, r \mapsto pr_2\left({}_{\beta}\tau \circ {}_{\theta}\tau^{-1}(x,r)\right)$ is a linear homeomorphism of R onto itself, and $x \mapsto \left(r \mapsto \left(D\left(\beta \circ \theta^{-1}\right)(\theta(x))\right)(r)\right)$ is a C^{k-1} morphism of $U \cap V$ into $L(R,R)$. It follows from (7.6) that $\left(\tau_X, T(X), X\right)$ is a vector bundle with the framing $\{(\theta, {}_{\theta}\tau),(U,\theta) \in \wedge\}$ and is called the tangent bundle over X. $T(X)$ is a C^{k-1} manifold with the atlas $\left\{\left(\tau_X^{-1}[U],(\theta \times id_R) \circ {}_{\theta}\tau\right);(U,\theta) \in \wedge\right\}$ modelled on $R \times R$ for each (U,θ) in $\wedge$. $(\theta \times id_R) \circ {}_{\theta}\tau$ will be denoted by $\underset{\sim}{\theta}$ and the frame $(\theta, {}_{\theta}\tau)$ is said to be induced by θ.

(8.2) Let X and Y be C^k manifolds modelled on Banach spaces R and S over A respectively, and let f be a C^k manifold morphism of X into Y. Denote $\underset{x \in X}{U} T_x f$ by $T(f)$ or Tf, Then $T(f)$ is a map of $T(X)$ into $T(Y)$ such that for each v in $T(X)$, $(T(f))(v) = \left(T_{\tau_X(v)} f\right)(v)$ that is for each x in X and each v in $T_x X$, $(Tf)(v) = (T_x f)(v)$

Let θ be a chart for X and let β be a chart for Y, such that $f[Dm(\theta)] \subset Dm(\beta)$. Then by (3.3) the local representative of Tf relative to the chart $(\theta \times id_R) \circ {}_{\theta}\tau$ for TX and the chart $(\beta \times id_S) \circ {}_{\beta}\tau$ for TY is the C^{k-1} manifold morphism,

$$ (\beta \times id_S) \circ {}_{\beta}\tau \circ Tf \circ \left((\theta \times id_R) \circ {}_{\theta}\tau\right)^{-1} : Im(\theta) \times R \to Im(\beta) \times S $$

given by

$$(\beta \times \mathrm{id}_S) \circ {}_{\beta}\tau \circ \mathrm{Tf} \circ \left({}_{\theta}\tau^{-1} \circ (\theta^{-1} \times \mathrm{id}_R)\right)(r', r) \text{ for: }$$

$$(r', r) \in \mathrm{Im}(\theta) \times R$$

$$= (\beta \times id_S) \circ {}_{\beta}\tau \circ Tf\left({}_{\theta}\tau^{-1}\left(\theta^{-1}(r'), r\right)\right)$$

$$= (\beta \times id_S) \circ_{\beta} \tau \circ Tf(x, \underline{c})$$

where $\theta^{-1}(r') = x$ and $r = D(\theta \circ c)(0)$, by the notation of (8.1),

$$= (\beta \times \mathrm{id}_S) \circ_{\beta} \tau\left(f(x), \underline{f \circ c}\right)$$

$$= (\beta \times id_S)\left(f(x), D(\beta \circ f \circ c)(0)\right)$$

$$= \left(\beta(f(x)), D(\beta \circ f \circ c)(0)\right)$$

$$= \left(\beta\left(f\left(\theta^{-1}(r')\right)\right), D\left(\beta \circ f \circ \theta^{-1} \circ \theta \circ c\right)(0)\right)$$

$$= \left(\beta \circ f \circ \theta^{-1}(r'), D\left(\beta \circ f \circ \theta^{-1}\right)(\theta(x)) \circ D(\theta \circ c)(0)\right)$$

$$= \left(\beta \circ f \circ \theta^{-1}(r'), D\left(\beta \circ f \circ \theta^{-1}\right)(\theta(x))(r)\right)$$

$$= \left(\beta \circ f \circ \theta^{-1}(r'), D\left(\beta \circ f \circ \theta^{-1}\right)(r')(r)\right)$$

hence Tf is a manifold morphism of $T(X)$ into $T(Y)$ called the tangent map of f.

Clearly $f \circ \tau_X = \tau_Y \circ (Tf)$, (8.2.1)

$$\begin{array}{ccc} TX & \xrightarrow{\tau_X} & X \\ {\scriptstyle Tf}\downarrow & & \downarrow{\scriptstyle f} \\ TY & \xrightarrow{\tau_Y} & Y \end{array}$$

L.H.S. $f \circ \tau_X(v) = f(x)$,

R.H.S. $\tau_Y \circ Tf(v) = \tau_Y\left(f(x), \underline{f \circ c}\right) = f(x)$ where $v = (x, \underline{c})$

Hence (f, Tf) is a vector bundle morphism of $(\tau_X, T(X), X)$ into (τ_Y, TY, Y).

Evidently - $T(id_X) = id_{T(X)}$

Let g be a C^k manifold morphism of Y into a C^k manifold Z. Then

$$T(g \circ f) = Tg \circ Tf \tag{8.2.2}$$

Evidently for any chart θ for X, $(\theta, T\theta)$ is a C^k vector bundle isomorphism of the tangent bundle $(\tau_{Dm(\theta)}, T(Dm(\theta)), Dm(\theta))$ over $Dm(\theta)$ onto the tangent bundle $(\tau_{Im(\theta)}, T(Im(\theta)), Im(\theta))$ over $Im(\theta)$. For each v in $T(Dm(\theta))$.

$$((\theta \times id_R) \circ {}_\theta\tau)(v) = (\theta(\tau_X(v)), d_{\tau_X(v)}\theta(v)) \text{ and}$$

$$(T\theta)(v) = \left(\theta(\tau_X(v), T_{\tau_X(v)}\theta(v))\right). \tag{8.2.3}$$

In view of this $(\theta \times id_S) \circ {}_\theta\tau$ will be identified with the tangent map $T\theta$ of θ, $T\theta$ is a chart for TX, said to be induced by θ.

(8.3) For any two charts θ and β for X, $T(\beta \circ \theta^{-1})$ is the function defined by

$$\theta[Dm(\theta) \cap Dm(\beta)] \times R \ni (r_1, r_2) \mapsto \left(\beta \circ \theta^{-1}(r_1), \left(D(\beta \circ \theta^{-1})(r_1)\right)(r_2)\right)$$

$$\in \beta[Dm(\theta) \cap Dm(\beta)] \times R.$$

(8.3) Let X and Y be open subset of Banach space R and S over A respectively, and let f be a C^k manifold morphism of X into Y.

For each x in X, $T_x X$ is identified with R in view of the linear homeomorphism τ_X of $T_x X$ into R and with $\{x\} \times R$ by means of the isomorphism $T_x X \ni v \mapsto (x, \tau_X(v)) \in \{x\} \times R$.

Let ${}_\tau X$ be the function on TX into $X \times R$ suchthat for each x in X and each v in $T_x X$, $\tau^X(v) = (x, \tau_X(v))$, Let τ^Y be the function on TY into $Y \times S$ defined similarly as above. Then τ^X is an isomorphism of the tangent bundle (τ_X, TX, X) with the trivial bundle $(pr_1, X \times R, X)$ relative to X, called canonical, (τ_X, TX, X) will be often identified with $(pr_1, X\times R, X)$. The tangent map Tf of f will be identified with the map $\hat{f}: X\times R \to Y\times S$,

$$(x,r)\mapsto(f(x), Df(x)\cdot r) \tag{8.3.1}$$

then $\hat{f}\circ\tau^X = \tau^Y\circ Tf$. By the same way $T(T(X))$ will be identified with $(X\times R)\times(R\times R)$, and the map $T(Tf)$ of $T(T(X))$ into $T(T(Y))$ will be identified with the map is

$$((x,r_2),(r_3,r_4))\mapsto\left((f(x), Df(x)\cdot r_2),\left(Df(x)\cdot r_3, D^2f(x)\cdot(r_2,r_3)+Df(x)\cdot r_4\right)\right)$$

as shown below

$$X\times R\times R\times R \xrightarrow{T(Tf)} (Y\times S)\times S\times S.$$

Take an element $((x,r_2),(r_3,r_4))\in(X\times R)\times R\times R$, then

$$\begin{aligned}
&T(Tf)((x,r_2),(r_3,r_4))\\
&=(Tf(x,r_2), DTf(x,r_2).(r_3,r_4))\\
&=\left((f(x), Df(x).r_2),\left(Df(x).r_3, D((x,r_2)\to Df(x).r_2)(r_3,r_4)\right)\right)
\end{aligned}$$

Now we have $(x,r_2)\mapsto(f(x), Df(x)\cdot r_2)$

consider $(x,r_2)\xrightarrow{h}(Df(x),r_2)\xrightarrow{k}Df(x).r_2$

$$\begin{aligned}
D(k\circ h)(x,r_2).(r_3,r_4) &= Dk(h(x,r_2))\circ Dh(x,r_2).(r_3,r_4)\\
&= Dk(Df(x),r_2)\circ\left(D^2f(x).r_3,r_4\right)\\
&= k(Df(x),r_4)+k\left(D^2f(x).r_3,r_2\right)\\
&= Df(x)\,r_4+D^2f(x).(r_3,r_2)\\
&= Df(x).r_4+D^2\,f(x).(r_2,r_3)
\end{aligned}$$

If $Y = TX$, which is identified with $X\times R$, then TY is identified with $(X\times R)\times R\times R$ and for each point $((x,r_2),(r_3,r_4))$ in $(X\times R)\times R\times R$, $(T(\tau_X))((x,r_2),(r_3,r_4)) = (x,r_3)$.

If we consider the trivial tangent bundle $(pr_1, X\times S, X)$, then $T(pr_1): T(X\times S)\to TX$, i.e., $T(pr_1): X\times S\times R\times S\to X\times R$.

Let $(x,s),(r,s')$ be any point of $(X\times S)\times R\times S$, then $T(pr_1)((x,s),(r,s'))$

$$= (pr_1(x,s), Dpr_1(x,s).(r,s'))$$
$$= (x, pr_1(r,s')) = (x,r) \text{ and}$$
$$(\tau_{X\times S})((x,s),(r,s')) = (x,s).$$

(8.4) Let (η, Y, X) be a C^k vector bundle over the C^k manifold X modelled on a Banach space R over A of fibre type a Banach space S over A. Then working by means of frames and using (8.3), we find that $(T(\eta), T(Y), T(X))$ is a C^{k-1} vector bundle over $T(X)$ of fibre type $S\times S$, and $(\tau_Y, T(Y), Y)$ is the C^{k-1} tangent bundle over Y of fibre type $R\times S$.

Now suppose that $Y = T(X)$ and $\eta = \tau_X$. Then (η, Y, X) is the tangent bundle $(\tau_X, T(X), X)$ over X. Hence $(\tau_{T(X)}, T(T(X)), T(X))$ is the tangent bundle over $T(X)$, quite different from the first.

Evidently $\tau_X \circ (\tau_{T(X)}) = \tau_X \circ T(\tau_X)$, for any two charts θ and β for X, $T\theta$ and $T\beta$ are the charts for TX induced by θ and β respectively and so $T(T\theta)$ and $T(T\beta)$ are charts for $T(TX)$ and for each points (r_1, r_2, r_3, r_4) in $\theta[Dm(\theta)\cap Dm(\beta)]\times R\times R\times R$

$$\left(T\left(T\left(\beta\circ\theta^{-1}\right)\right)\right)(r_1, r_2, r_3, r_4)$$
$$= (\beta\circ\theta^{-1}(r_1), D(\beta\circ\theta^{-1})(r_1).r_2, D(\beta\circ\theta^{-1})(r_1).r_3,$$
$$D^2(\beta\circ\theta^{-1})(r_1).(r_2, r_3) + D(\beta\circ\theta^{-1})(r_1).r_4)$$

(8.5) Let X be a C^k manifold modelled on A^n, and let θ be any chart for X, then $\underset{\sim}{\theta} = (\theta\times id_{A^n})\circ {}_\theta\tau$ is the chart for TX associated with θ. Then for each v in $T(Dm(\theta))$ $\underset{\sim}{\theta}(v) = (\theta\times id_{A^n})\circ {}_\theta\tau(v)$

$$= (\theta\times id_{A^n})(x, \theta_x(v))$$
$$= (\theta(x), \theta_x(v))$$
$$= (\theta(\tau_X(v)), \theta_{\tau_X(v)}(v))$$

$$= \left(\theta(\tau_X(v)), \left(d_{\tau_X(v)}\theta\right)(v)\right)$$

$$= (\theta(x), (d_x\theta)(v)).$$

Hence,

$$(T\theta)^1(v) = \underset{\sim}{\theta}^1(v) = \theta^1(\tau_X(v)), \ldots, (T\theta)^n(v)$$

$$= \underset{\sim}{\theta}^n(v) = \theta^n(\tau_x(v))$$

$$= (T\theta)^{n+1}(v) = \underset{\sim}{\theta}^{n+1}(v)$$

$$= \left(d_{\tau_X(v)}\theta^1\right)(v), \ldots, (T\theta)^{n+n}(v)$$

$$= \underset{\sim}{\theta}^{2n}(v) = \left(d_{\tau_X(v)}\theta^n\right)(v)$$

(8.6) Let X be a C^{k-1} $(k \in N)$ manifold modelled on a Banach space R over A. Let S be any Banach space over A. Let i and j be any natural number.

Then the C^k vector bundle $\left(T_j^i(\tau_X), T_j^i(T(X), S), X\right)$ is called the vector bundle over X of tensors, contravariant of order i and covariant of order j, or of type (i, j), with values in S, and a C^k section of this bundle is called a tensor field on X contravariant of order i and covariant of order j, or of type (i, j), with values in S. Then a C^k tensor field v on X of type (i, j) is a C^k morphism of X into $T_j^i(T(X), S)$ such that for each x in X, $v(x)$ is an $(i+j)$ - multilinear map of $\left((T_xX)^*\right)^i \times (T_xX)^j$ into S. If $S = A$ then the phrase 'with values in S' in the above definition will be deleted. A C^k tensor field on X of type (0,0) is called a scalar field on X. Every C^k morphism of X into A will thus be called a scalar field on X. We denote $T_1^0(TX, A)$ by T^*X and $T_1^0(\tau_X)$ by τ_X^*.

Let θ be a chart for X, then $(\theta, {}_\theta\tau)$ is a frame for TX. By (8.1.) $\left(\theta, T_j^i({}_\theta\tau)\right)$ is a frame and $\left(\theta \times id_{T_j^i(R,S)}\right) \circ T_j^i({}_\theta\tau)$ is a chart for $T_j^i(TX, R)$, said to be induced by θ we denote $T_1(R) = R^*$ and $(\theta \times id_R) \circ T_1({}_\theta\tau)$ by $(T\theta)^*$.

9

Vector Fields

(9.0) A (differentiable) vector field on R^2 is a (differentiable) function from R^2 to R^2, in other words, a vector field is a vector with coefficients that may vary for example in the plane $\underline{R}^2$, let (r, p) are the cartesian coordinate and (ρ , θ) are the polar coordinate then in the vector field $(-p, r)^T$ or $\frac{\partial}{\partial\theta} = -p\frac{\partial}{\partial r} + r\frac{\partial}{\partial p}$ where

$\frac{\partial}{\partial r}, \frac{\partial}{\partial p}$ are the basis vector in $\underline{R}^2$, the coefficients -p and r do indeed vary, they are not constant as function of r and p, also

$$\frac{\partial}{\partial\rho}\begin{pmatrix} r/\sqrt{r^2+p^2} \\ p/\sqrt{r^2+p^2} \end{pmatrix} = \frac{1}{\sqrt{r^2+p^2}}\left(r\frac{\partial}{\partial r} + p\frac{\partial}{\partial p} \right)$$

is a vector field not defined on $\underline{R}^2$ because $\frac{\partial}{\partial\rho}$ is not defined at the origin it is defined only for $\rho>0$. So we define A differentiable vector field on an open subset U of $\underline{R}^2$ and say that a differentiable vector field on U is a differentiable function from U to $\underline{R}^2$.

It is not hard to define vector field on any differentiable manifold. The only issue is compatibility accross coordinate systems. For instance, to define a vector field on a manifold X we can specify a coordinate vector field in each coordinate patch of X and require that two coordinate vector fields should agree on the overlap of the coordinate patch.

(9.1) In the theory of Lie groups and transformation groups the main tool is the "infinitesimal transformation". In order to present this, we need first to develop the concept of a vector field on a manifold. Suppose c is a smooth curve on a manifold X, Parametrized by $c : I \to X$, where I is an open interval of **R**. In local coordinate $x = (x^1, ..., x^n)$, c is given by n smooth functions $c(t) = (c^1(t), ..., c^n(t))$ of the real variable t. At each point $x = c(t)$ of X the curve has a tangent vector namely the derivative $\dot{c}(t) = dc/dt = (\dot{c}^1(t), \ldots, \dot{c}^n(t))$. In order to distinguish between tangent vectors and coordinate expressions for points on the manifold, we adopt the notation

$$v|_x = \dot{c}(t) = \dot{c}^1(t)\frac{\partial}{\partial x^1} + \dot{c}^2(t)\frac{\partial}{\partial x^2} + \ldots + \dot{c}^n(t)\frac{\partial}{\partial x^n} \qquad (9.1.1)$$

for the tangent vector to c at $x = c(t)$. For example, the helix

$$c(t) = (\cos t, \sin t, t) \qquad (9.1.2)$$

in R^3, with coordinates x, y, z, has tangent vector

$\dot{c}(t) = -\sin t\frac{\partial}{\partial \mathbf{x}} + \cos t\frac{\partial}{\partial \mathbf{y}} + \frac{\partial}{\partial \mathbf{z}} = -\mathbf{y}\frac{\partial}{\partial \mathbf{x}} + \mathbf{x}\frac{\partial}{\partial \mathbf{y}} + \frac{\partial}{\partial \mathbf{z}}$ at the point

$(x, y, z) = c(t) = (\cos t, \sin t, t)$. Two curves $C = \{c(t)\}$ and $\overline{C} = \{\overline{c}(\overline{t})\}$ for some t, $\overline{t}$ passing through the same point $x = c(t_0) = \overline{c}(\overline{t_0})$ for some t_0, $\overline{t_0}$, have the same tangent vector if and only if their derivatives agree at the point $\frac{dc}{dt}(t) = \frac{d\bar{c}}{dt}(\bar{t})$ (9.1.3)

The concept defined above is independent of the local coordinate system used near x, Indeed, if $x = c(t) = (c^1(t), ..., c^n(t))$ in the local coordinate expression in terms of $x = (x^1, ..., x^n)$ and $y = f(x)$ is any diffeomorphism, then $y = f(c(t))$ is the local coordinate formula for the curve in terms of the y-coordinates. The tangent vector $v|_x = \dot{c}(t)$ which has the formula (9.1.1) in the x-coordinates, takes the form

$$v|_{y=f(x)} = \sum_{j=1}^{n} \frac{d}{dt} f^j(c(t)) \frac{\partial}{\partial y^j} = \sum_{j=1}^{n} \sum_{k=1}^{n} \frac{\partial f^j}{\partial x^k}(c(t)) \frac{dc^k}{dt} \frac{\partial}{\partial y^j} \qquad (9.1.4)$$

in the y-coordinates. Since the Jacobian matrix $\frac{\partial f^j}{\partial x^k}$ is invertible at each point, (9.1.3) holds if and only if

$$\frac{d}{dt} f(c(t)) = \frac{d}{d\bar{t}} f(\bar{c}(\bar{t}))$$

Note that (9.1.4) tells us that how a tangent vector (9.1.1) behaves under the given change of coordinates $y = f(x)$.

As given in chapter 3, the collection of all tangent vectors to all possible curves passing through a given point x in X is called the tangent space to X at x, and is denoted by $T_x X$. If X is an n-dimensional manifold, then $T_x X$ is an n-dimensional vector space, with $\left\{ \frac{\partial}{\partial x^1}, ..., \frac{\partial}{\partial x^n} \right\}$ as a basis in the given local coordinates. The collections of all tangent spaces corresponding to all points x in X is called the tangent bundle of X, denoted by $TX = \bigcup_{x \in X} T_x X$.

These tangent spaces are "glued" together in an obvious smooth fashion, so that if c(t) is any smooth curve then the tangent vectors

$\dot{c}(t) \in T_{c(t)}X$ will vary smoothly from point to point. This makes the tangent bundle TX into a smooth manifold of dimension 2n.

For example, if $X = R^n$, then we can identify the tangent space T_xR^n at any $x \in R^n$ with R^n itself. This stems from the fact that the tangent vector $\dot{c}(t)$ to a smooth curve c(t) can be realized as an actual vector in R^n, namely $(\dot{c}^1(t),\ldots,\dot{c}^n(t))$. Another way of looking at this identification is that we are identifying the basis vector $\frac{\partial}{\partial x^i}$ of T_xR^n with the standard basis vector e_i of R^n. The tangent bundle of R^n is thus a cartesian product $TR^n \cong R^n \times R^n$. If S is a surface in R^3, than the tangent space T_xS can be identified with the usual geometric tangent plane to S at each point $x \in S$. This again uses the identification $T_xR^3 \cong R^3$, so $T_xS \subset T_xR^3$ is a plane in R^3.

A vector field v on X assigns a tangent vector $v(x) \in T_xX$ to each point $x \in X$, with v(x) varying smoothly from point to point. In local coordinates $(x^1,\ldots,x^n)$, a vector field has the form.

$$v|_x = \xi^1(x)\frac{\partial}{\partial x^1} + \xi^2(x)\frac{\partial}{\partial x^2} + \ldots + \xi^n(x)\frac{\partial}{\partial x^n}$$

where each $\xi^i(x)$ is a smooth function of x and $\left\{\frac{\partial}{\partial x^i}\right\}$ is the basis associate to coordinate system $(x^1,\ldots,x^n)$.It is clear that v is differentiable if and only if the functions ξ^i are differentiable for some coordinate system. A good physical example of a vector field is the velocity field of a steady fluid flow in some open subset $X \subset R^3$. At each point $p(x, y, z) \in X$, the vector v (p) would be the velocity of the fluid particle passing through the point (x, y, z).

An integral curve of a vector field is a smooth parameterized curve x = c(t) whose tangent vector at any point coincides with the value of v at the same point: $\dot{c}(t) = v|_{c(t)}$ for all t. In local coordinate,

$$x = c(t) = (c^1(t), ..., c^n(t))$$

must be a solution to the autonomous system of ordinary differential equations

$$\frac{dx^i}{dt} = \xi^i(x)\ , i=1,...,n \tag{9.1.5}$$

where the $\xi^i(x)$ are the coefficients of v at x. For $\xi^i(x)$ smooth, the standard existance and uniqueness theorems for systems of ordinary differential equations guarantee that there is a unique solution to (9.1.5) for each set of initial value

$$c(0) = x_0 \tag{9.1.6}$$

which implies that the existence of a unique maximal integral curve $c: I \to X$ passing through a given point $x_0 = c(0) \in X$, where maximal means that it is not contained in any longer integral curve, i.e., if $\bar{c}: \bar{I} \to X$ is any other integral curve with some initial value $\bar{c}(0) = x_0$, then $\bar{I} \subset I$ and $\bar{c}(t) = c(t)$ for $t \in \bar{I}$. Note that if $v|_{x_0}$, then the integral curve through x_0 is just the point $c(t) = x_0$ itself, defined for all t.

If $x = (x^1, ..., x^n)$ is a local coordinate system on an open set U of X, then at each point x of U, v(x) can be expressed uniquely as-

$$v(x) = \sum_{i=1}^{n} \xi^i(x) \left(\frac{\partial}{\partial x^i} \right)_x \tag{9.1.7}$$

Then n functions $\xi^i (i = 1,...,n)$ on U are called the component of v with respect to the local coordinate system $(x^1,...,x^n)$. If $(\bar{x}^1,...,\bar{x}^n)$ is another local coordinate system on U, and $\bar{\xi}^i$ the components of v with respect to $(\bar{x}^1,...,\bar{x}^n)$, then we have

$$\bar{\xi}^i(x) = \sum_{j=1}^{n} \frac{\partial \bar{x}^i}{\partial x^j}(x)\xi^j(x),\ x \in U \tag{9.1.8}$$

Thus, the property that the components of v are continuous or of class c^r at one point does not depend on the choice of the local coordinate system. If all the components of v are continuous or of class c^r on U, then we say that v is continuous of class c^r on U. if v is continous or of class c^r on a neighbourhood of each point of X, than v is said to be continuous or of class c^r on X. We will consider c^∞ vector fields and so denote by $\Gamma^\infty(X)$ the set of all c^∞ vector fields on X.

If $(x^1,...,x^n)$ a local coordinate system on U, we can define a vector field $\partial/\partial x^i$ on U by $x \to (\partial/\partial x^i)_x$. The components ξ^i of a vector field with respect to $(x^1,...,x^n)$ are c^∞ functions on U, and v can be cxpressed on U as

$$v = \sum_{i=1}^{n} \xi^i \frac{\partial}{\partial x^i}$$

(9.2) Let X be a c^k ($k \in N$) manifold modelled on a Banach space S over A. Then a $c^{k'}(k' \in N, k' \le k)$ section of the tangent bundle τ_x of X is called a $c^{k'}$ tangent vector field or simply vector field on X. Thus a $c^{k'}$ vector fields v on X is a $c^{k'}$ morphism of X into TX such that, for each x in X, v(x) is in T_xX. The set of all $c^{k'}$ vector fields on X will be denoted by $\Gamma^{k'}(X)$. If X is an open subset of S then a $c^{k'}$ vector field v on X will be taken as a $c^{k'}$ morphism of X onto S i.e., $X \xrightarrow{v} T_xX \xrightarrow{\tau_x} S$ given by $\mathbf{X \ni x \mapsto \tau_x(v(x)) \in S}$

Let v be a $c^{k'}$ vector field on X let θ be any chart for X. then for any vector s in S, the function

$$x \mapsto (d_x\theta)^{-1}(s) \tag{9.2.1}$$

is a c^{k-1} vector field on the open submanifold Dm(θ) of X and will be denoted by S_θ. If X is an open subset of S than for each x in X, τ_x $(S_\theta(x))$ = s, where θ = id_x. Also the function

$Im(\theta) \ni s \mapsto (d_{\theta^{-1}(s)}\theta)(v(\theta^{-1}(s))) \in S$ is called the vector part of v and is denoted by V_θ. Now let α be another chart for X such that $Dm(\alpha) = Dm(\theta)$. Then for each s in S.

$$V_\alpha(s) = (D\,(\alpha o \theta^{-1})\,((\alpha o \theta^{-1})^{-1}(s))\,(v_\theta\,((\alpha o \theta^{-1})^{-1}\,(s))) \quad (9.2.2)$$

(9.3) Let $(g_i,\ i \in I)$ be a basis of S. Then for each $i \in \underline{n}$. $g_{i\,\theta}$ is called the ith basic vector field relative to θ, if X is an open subset of $\underline{R}^n$ then $g_{i(id_x)}$ is called the ith - cononical basic vector field on X and is denoted by $\bar{g}_i$.

(9.4) Let X be a c^k manifold modelled on $\underline{R}^n$. Let Y be a c^k manifold modelled on R^p and let m be a c^k manifold morphism of X into Y. Let θ and ϕ be charts for X and Y respectively such that $m[Dm(\theta)] \subset Dm(\phi)$. Let v be a c^{k-1} vector field on X, let $(g_i : i \in \underline{n})$ be the cononical basis of $\underline{R}^n$ and let $(h_j : j \in p)$ be the canonical basis of $\underline{R}^p$. Then for each x in $Dm(\theta)$.

$$v(x) = \sum_{i \in \underline{n}} dx^i\,(v(x))\ \bar{g}_i\,(x) \quad (9.4.1)$$

$$Tm \circ v(x) = \sum_{j \in \underline{p}} d_x(\phi^j \circ m)(v(x))\bar{h}_j(m(x)) \quad (9.4.2)$$

(9.5) Let X be a c^k $(k \in N)$ manifold modelled on a Banach space S over A. let c be a $c^{k'}$ $(0 \leq k' \leq k)$ morphism of an open interval J of R with center q into a $c^{k'}$ manifold X. Then Tc is the tangent map of the tangent bundle $TJ \cong J \times R$ into the tangent bundle TX. Let $\bar{1}$ be the vector field on J such that for each t in J, $\bar{1}(t) = (\tau^J)^{-1}(t, 1)$ called the basic vector field on J, where τ^J is the canonical isomorphism of TJ onto $J \times R$ defined in (8.3). The $c^{k'-1}$ morphism $Tc \circ \bar{1}$ of J into TX will be denoted by $\dot{c}$. For each t in J, $\dot{c}(t)$ is called the velocity vector of c at t and is denoted by dc(t)/dt. If X is an open subset of S, then

$$dc(t)/dt = Dc\ (t).\ 1 = \tau_{c(t)}\ (Tc \circ \bar{1}(t)) = \tau_{c(t)}(\dot{c}(t))$$

Let v be a $c^{k'}$ $(0 \leq k' \leq k)$ vector field on X, then by an integral solution for v is meant a morphism c of J into X such that $\dot{c}(t) = dc/dt = v \circ c(t)$ for all t in J, called the differential equation defined by v. By an integral solution for v at a point x of X is meant a $c^{k'+1}$ morphism of an open interval J with centre q of R into X such that for each t in J, $\dot{c}(t) = v(c(t))$ and $c(q) = x$.

We also say that an integral curve of a vector field v is a smooth parametrized curve $x = c(t)$ whose tangent vector at any point coincides with the value of v at the same point: $\dot{c}(t) = v(c(t))$ for all t. In local coordinates $x = c(t) = (c^1(t), \ldots, c^n(t))$ must be a solution to the autonomous system of ordinary differential equations $dx^i/dt = \xi^i(x)$, $(i = 1, \ldots, n)$, where $\xi^i(x)$ are the coefficient of v at x, without loss of generality we may take $q = 0$.

(9.6) Let v be a c^k vector field on a c^γ manifold X, let x be any point of X. then the union of all integral solution for v at x is also on integral solution for v at x, called the maximal integral solution for v at x, we denote the domain of the maximal integral solution for v at x by J_x, J_x is an open interval of R.

We note that if v is any smooth vector field on a manifold X and f(x) is any smooth real valued function defined for $x \in X$, then f.v is again smooth vector field with $(f.v)(x) = f(x)\, v(x)$. In local coordinates, if $v = \sum_{i=1}^{n} \xi^i(x) \frac{\partial}{\partial x^i}$ then $f \cdot v = \sum_{i=1}^{n} f(x)\xi^i(x) \frac{\partial}{\partial x^i}$.

If f never vanishes, the integral curves of f.v coincide with the integral curve of v, but the parametizations will differ, for instance, the integral curves for 2v will be traversed twice as fast as those of v, but otherwise will be the same subset of X.

(9.7) *Flows:*

If v is a vector field, we denote the parametrized maximal integral curve passing through x in X by f(t, x) and call f the flow generated by v. Thus by a flow for v is meant a morphism f of $J \times U$ into X where J is an open interval of $\underline{R}$ with center 0 and U is an open neighbourhood of x contained in X such that for each u in U the map $f_{\cdot u} : J \to X$ defined by

$$J \ni t \mapsto f(t, u) \in X \tag{9.7.1}$$

is an integral solution for v at u, i.e., f (t, u) will be a point on the integral curve passing through u in X. A morphism f of $J \times U$ into X where J is an open interval of $\underline{R}$ with center 0 and U is an open neighbourhood of u contained in X is a flow for v at u iff for each u in U, f (0, u) = u and for each point (t, u) in $J \times U$,

$$T(f \circ i_u) \circ \bar{1}\,(t) = \frac{df_{\cdot u}}{dt} = v \circ f \circ i_u\,(t) = v \circ f(t, u), \tag{9.7.2}$$

where i_u is the map.

$$J \ni t \mapsto (t, u) \to J \times U \tag{9.7.3}$$

(9.8) Let f be the flow for a c^∞ vector field v on a separated real c^∞ manifold X modelled on a Banach space. Then for each x in X and each s in J_x

$$J_x\text{-}s = J_{f(s,x)}$$

and for each t in $J_{f(s,x)}$

$$f(t,f(s,x)) = f(t+s,x)$$

Proof: Let x be any point of x, and let s be any point of J_x. Then $t \mapsto f(t,f(s,x))$ and $t \mapsto f(t+s,x)$ are both maximal integral solution for v at f(s,x) and as such have identical domain. Hence t is in $J_{f(s,x)}$ iff t+s is in J_x. Thus $J_{f(s,x)} = J_x$-s. By same token for all t in J_x-s, $f(t,f(s,x)) = f(t+s,x)$.

(9.9) Let X be a separated c^∞ manifold, v be a c^∞ vector field on X and let f be a c^∞ morphism of an open subset W of the product manifold $\underline{R} \times X$ into X such that f is the flow for v. For each t in $\underline{R}$

let f_t be the map $x \mapsto f(t, x)$ of X into it self. Then f is called a c^∞ action of $\underline{R}$ on X iff $f_0 = id_X$ and for each pair of elements r and s of R, $f_{r+s} = f_r \circ f_s$. Then for each t in $\underline{R}$ $f_{-t} = f_t^{-1}$, and so f_t is a c^∞ diffeomorphism of X onto itself. In view of this a c^∞ action of $\underline{R}$ on X is also called a one parameter group of c^∞ diffeomorphism of X.

Let f be a one parameter group of c^∞ diffeomorphism of X. Then the function v on X, which assigns to each x in X the value

$$v(x) = (Tf_{\cdot x})(\bar{1}(o)) = \left(\frac{df_{\cdot x}}{dt}\right)_{t=u} = (\dot{f}_{\cdot x})(o)$$

is a c^∞ vector field on X, said to be induced by f, f is the flow for the vector field v on X induced by it and v is complete.

On the otherhand let v be a complete c^∞ vector field on X. Then the flow for v is a one parameter group of c^∞ diffeomorphism of X and v is induced by it.

(9.9) (a) Let m be a c^∞ diffeomorphism of Xonto itself. Let g be the morphism of $W \subset \underline{R} \times X$ into X such that for each $(t,x) \in W$, $g(t,x) = f(t, m^{-1}(x))$. Then $m \circ g$ is the flow for the vector field $Tm \circ v \circ m^{-1}$ on X.

(b) A c^∞ vector field v on a separated c^∞ manifold X is said to be complete or globally integrable iff the domain of the flow for v is $\underline{R} \times X$.

A c^κ function $f: \underline{R} \times X \to X$ is called an enternal c^κ motion in X or a one parameter group of c^κ diffeomorphism of X iff for each t in $\underline{R}$, f_t is in $Diff^k(X)$,

The flow of a vector field has the following basic properties:

(c) $f(s, f(t, x)) = f(s + t, x)$, (9.9.1)

$x \in X$ and for all $s, t \in R$ such that both sides of equation are defined,

$$f(0, x) = x \tag{9.9.2}$$

and $$\frac{d}{dt} f(t,x) = v(f(t,x)) \tag{9.9.3}$$

for all t where defined. Here (9.9.3) simply states, that v is tangent to the curve f (t, x) for fixed x and (9.9.2) gives the initial conditions for this integral curve. Comparing the first two properties (9.9.1), (9.9.2) with the properties of local group action of Lie group ($\underline{R}$, +) on the manifold X, we see that flow generated by a vector field is the same as local group action of the Lie group $\underline{R}$ on the manifold X, often called a one parameter group of transformations. The vector field v is called the infinitesimal generator of the action since by Taylor's theorem, in local coordinates

$$f(t,x) = x + t\,\xi(x) + O(t^2) \qquad (9.9.4)$$

where $\xi = (\xi^1,\ldots,\xi^n)$ are the coefficient of v. The orbits of the one parameter group action are the maximal integral curves of the vector field v.Conversely, if f(t,x) is any one parameter group of transformations acting on X, then its infinitesimal generator is obtained by specializing (9.9.3) at t = 0.

$$v|_x = \left(\frac{d}{dt} f(t,x)\right)_{t=0} \qquad (9.9.5)$$

uniqueness of solutions to (9.9.5), (9.9.6) guarantees that the flow generated by v coincides with the given local action of $\underline{R}$ on X on the common domain of definition. Thus there is a one to one correspondance between local one parameter groups of transformations and their infinitesimal generators. The computation of the flow or one-parameter group generated by a given vector field v is often referred to as exponentiation of the vector field ,we use the notation.

$$\exp(tv)\,x \equiv f(t, x) \qquad (9.9.6)$$

for this flow here. In terms of this exponential notation, the properties (9.9.1), (9.9.2) and (9.9.3) can be restated as

$$\exp[(s+t)\, v]\, x = \exp(s\, v) \exp(tv)x \tag{9.9.7}$$

$$\text{whenever defined } \exp(0v) = x \tag{9.9.8}$$

and $\frac{d}{dt}[\exp(t\, v)\, x] = v|(\exp(t\, v)\, x$ for all $x \in X$.

(9.9.9)

In particular, $v|_x$ is obtained by evaluating (9.9.9) at $t = 0$.

(9.10) *Example:*

1. Let $X = R$ with coordinate x, and consider the vector field $v = \frac{\partial}{\partial x} \equiv \partial_x$. Then $\exp(t\, v)\, x = \exp(t\, \partial_x)\, x = (1 + t\, \partial_x + \ldots)\, x = x + t$,

   which is globally defined. For the vector field $x\, \partial_x$ we have the usual exponential

   $\exp(t\, x\, \partial_x)\, x = (1 + t\, x\, \partial_x + \frac{(tx\partial_x)^2}{2} + \ldots)x$

   $x + t\, x = (1 + t)\, x = e^t.x$

   since it must be the solution to the ordinary differential equation $\frac{dx}{dt} = x$ with initial value x at $t = 0$.

2. In the case of $\underline{R}^n$, a constant vector field $v_a = \sum a^i \frac{\partial}{\partial x^i}$, $a = (a^1, \ldots, a^n)$, exponentiates to the group of translations $\exp(t\, v_a)\, x = x + ta$, $x \in R^n$.

   In the direction a. similaryly, a linear vector field

   $$V_A = \sum_{i=1}^{n} \left(\sum_{j=1}^{n} a_{ij} x^j \right) \frac{\partial}{\partial x^i}$$

   where $A = (a_{ij})$ is an $n \times n$ matrix of constants, has flow $\exp(t\, v_A)x = e^{tA}\, x$

where $e^{tA} = 1 + tA + \frac{1}{2}t^2A^2 + \ldots$ is the usual matrix exponential.

3. Consider the group of rotations in the plane
$\psi(t,(x,y)) = (x \cos t - y \sin t,\ x \sin t + y \cos t)$
Its infinitesimal generator is a vector field $v = \xi(x,y)\,\partial_x + \eta(x,y)\,\partial_y$, where according to (9.9.5)

$$\xi(x,y) = \left.\frac{d}{dt}\right|_{t=0} (x\cos t - y\sin t) = -y$$

$$\eta(x,y) = \left.\frac{d}{dt}\right|_{t=0} (x\sin t + y\cos t) = x$$

Thus $v = -y\partial_x + x\partial_y$ is the infinitesimal generator and indeed, the above group transformation agree with the solutions to the system of ordinary differential equations.

$$\frac{dx}{dt} = -y, \qquad \frac{dy}{dt} = x$$

4. Consider the local group action occurs in the study of the heat equation
Let $X = R^2$, $G = R$ and consider the local group action ψ (t, (x, y)) $= \left(\frac{x}{1-tx}, \frac{y}{1-tx}\right)$, which is defined on U = {(t, (x, y)):

$$t < \frac{1}{x} \text{ for } x > 0, \text{ or } t > \frac{1}{x}, \text{ for } x < 0\} \subset \underline{R} \times \underline{R}^2$$

Differentiating, as before, we find the infinitesimal generator to be $v = x^2\partial_x + xy\partial_y$. This shows that a smooth vector field may still generate only a local group action.

(9.11) Let m be a c^k ($k \in N$) morphism of a separated real c^k manifold X into a separated c^k manifold Y, let v (respectively w) be a c^k ($k \in N$) vector field on X (respectively on Y). Then w is said to be m-related to v if $Tm \circ v = w \circ m$. If w is m-related to v then the function on m(X) into TY, which assigns to each y in m(X) as value the element (Tm) (v(x)) of TY, where x in X is such that y = m(x), will be denoted by $m_* v$; evidently $m_* v = w|m(X)$ and as such it is a vector field on m(X). Evidently for each x in X, (Tm) (v(x)) = $(m_* v)$ (m(x)).

Let f (respectively g) be a flow for v (respectively w) at a point x of X (at a point m(x) of Y), let U be any open neighbourhood of x in X and let V be on open neighbourhood of m(x) in Y and be J be on open interval about 0 in R such that $J \times U = Dm(f)$, $J \times V = Dm(g)$ and $m(U) \subset V$.

First suppose that for each t in J, $m \circ f_t = g_t \circ m$. Then for each t in J and each u in U, m(f(t,u)) = g(t, m(u))

and so $(Tm)((d(f_{\cdot u}(t))/dt) = (d(g_{\cdot m(u)}(t))/dt)$.

But $(d(f_{\cdot u}(t)))/dt = v(f(t,u)) = v(f_{t\cdot}(u))$

and $(d(g_{\cdot m(u)}(t)))/dt = w(g(t,m(u))) = w(g_{t\cdot}(m(u)))$.

hence (Tm (v (f (t, u)))) = w (m (f (t, u))),

i.e., $Tm \circ v \circ f_t = w \circ m \circ f_t$

Consequently $Tm \circ v = w \circ m$

Thus w is m-related to v.

Next suppose that w is m-related to v. Then $(Tm) \circ v = w \circ m$. Let u be any element of U. For each t in J let $F(t) = m(f_{t\cdot}(u))$ and $G(t) = g_{t\cdot}(m(u))$. Then F(0) = G(0) and for each t in J.

$$\begin{aligned} dF(t)/dt &= (Tm)((df_{\cdot u})(t)/dt) \\ &= (Tm)(v(f(t,u))) \\ &= w(m(f(t,u))) \\ &= w(m(f_{t\cdot}(u))) \\ &= w(F(t)) \end{aligned}$$

$$\begin{aligned} \text{and } dG(t)/dt &= (dg_{\cdot m(u)}(t))/dt \\ &= w(g(t, m(u))) \end{aligned}$$

$= w(g_{t}(m(u))$
$= w(G(t))$

since F and G satisfy the same differential equation and F(o) = G (o), for each t in J,

$$m \circ f_t = g_t \circ m$$

(9.12) Let f be a diffeomorphism of a separated smooth real manifold X onto itself. Let v be a vector field on X, let a morphism g of J × U into X be a flow for v at a point x of X, where J is an open interval of $\underline{R}$ with center 0 and U is an open neighbourhood of x contained in X, and let h be the function.

$$J \times f[U] \ni (t, z) \mapsto f(g(t, f^{-1}(z))) \in X.$$

Then h is a flow for the vector field $w = Tf \circ v \circ f^{-1}$ on X at f (x). If v = w then for each u in $U \cap f(U)$ and each t in J, $g(t, f^{-1}(u)) = f^{-1}(g(t,u))$

Proof: for each u in U let in i_u be the function $J \ni t \mapsto (t, u) \in J \times U$, since g is a flow for v at x, by (9.7) for each t in J, and each u in U

$T(g \circ i_u) \circ \bar{1}(t) = v \circ g(t, u)$.

For $z \in f(U)$, let J_z be the function $J \ni t \mapsto (t, z) \in J \times f(U)$ then $h \circ J_z = f \circ g \circ i_{f^{-1}(z)}$.

Hence for each t in J and each z in f (U)

$$T(h \circ J_z) \circ \bar{1}(t) = T(f \circ g \circ i_{f^{-1}(z)})(t,1)$$
$$= Tf \circ (T(g \circ i_{f^{-1}(z)}))(t,1)$$
$$= Tf \circ v \circ g(t, f^{-1}(z))$$
$$= Tf \circ v \circ f^{-1} \circ f \circ g(t, f^{-1}(z))$$
$$= w(h(t, z))$$

since g is a flow for v at x, g (0, x) = x.

Hence $h(0, f(x)) = f(g(0, f^{-1}(f(x)))$
$= f(g(0, x)) = f(x)$

Thus h is a flow for w at f (x).

Suppose that v = w then by the uniqueness of flow for v at x, for each t in J, and each u in $U \cap f(U)$, $h \circ J_u = f \circ g \circ i_{f^{-1}(u)}$

or $f^{-1} \circ h \circ J_u = f^{-1} \circ f \circ g \circ i_{f^{-1}(u)}$

or $f^{-1} \circ h \circ J_u(t) = g(i_{f^{-1}(u)})(t)$

$f^{-1}(h(t,u)) = g(t, f^{-1}(u))$, but g = h,

hence $f^{-1}(g(t,u)) = g(t, f^{-1}(u))$

(9.13) Let X be a c^k ($k \in N$) manifold modelled on a Banach space E over A, Let v be a $c^{k'}$ ($k' \in N, k' < k$) vector field on X. let W be an open subset of X. For each c^k morphism f of W into A, let $L_v(f)$ be the function on W into A such that for each x in W.

$$L_v(f)(x) = L_{(x,v(x))}(f) \tag{9.13.1}$$

the notation being that of (5.5). We write simply $L_v f$ for $L_v(f)$ and note that $L_v f$ is a $c^{k'}$ morphism of W in to A. For any two c^k morphism f and g of W into A and for any two scalars a and b in A.

$$L_v(af + bg) = aL_v f + bL_v g \tag{9.13.2}$$

$$L_v(fg) = (L_v f)g + f(L_v g) \tag{9.13.3}$$

If there is an x in W such that v (x) ≠ 0 then there is a c^k morphism f of W into A such that $L_v f \neq 0$.

If θ is a chart for X then for each r in E, each x in Dm (θ) and each c^k morphism f of Dm (θ) into A, then by the notion given in (9.2).

$$L_{S_\theta} f(x) = L_{(x,S_\theta(x))} f = (d_x f)(S_\theta(x)) \text{ (by (5.5.1))} \tag{9.13.4}$$

$$= (d_x f)(d_x \theta)^{-1}(r) \text{ (by using(9.2.1))}$$

$$= (D(f \circ \theta^{-1})(\theta(x)))(\theta_x)(d_x\theta)^{-1}(r)$$

$$= (D(f \circ \theta^{-1})(\theta(x))(r) \text{ by (5.4)}$$

$$= D_r(f \circ \theta^{-1})(\theta(x))$$

and so $(L_{S_\theta} f)(x)$ is the derivative of $f \circ \theta^{-1}$ at $\theta(x)$ along r.

If m is a c^k diffeomorphism of a c^k manifold X into a c^k manifold Y then for each c^k morphism f of X into A and each y in Y.

$$L_{(Tm\circ v\circ m^{-1})}(f\circ m^{-1})(y)$$

$$= (id_A)_{f(m^{-1}(y))} \circ (T_y(f\circ m^{-1}))(Tm\circ v\circ m^{-1}(y))$$

$$= (id_A)_{f(m^{-1}(y))} \circ T_{m^{-1}(y)} f\circ T_y m^{-1} \circ T_{m^{-1}(y)} m(v(m^{-1}(y)))$$

$$= (id_A)_{f(m^{-1}(y))} \circ T_{m^{-1}(y)} f\circ T_{m^{-1}(y)} (m^{-1}\circ m)(v(m^{-1}(y))$$

$$= (id_A)_{f(m^{-1}(y))} \circ T_{m^{-1}(y)} f\circ T_{m^{-1}(y)} (id_X)(v(m^{-1}(y))$$

$$= (id_A)_{f(m^{-1}(y))} \circ T_{m^{-1}(y)} f(v(m^{-1}(y))$$

$$= (d_{m^{-1}(y)} f)(v(m^{-1}(y))$$

$$= L_{(m^{-1}(y),v(m^{-1}(y)))} f$$

$$= L_v f(m^{-1}(y))$$

Hence $L_{Tm\circ v\circ m^{-1}}(f\circ m^{-1}) = L_v f\circ m^{-1}$ (9.13.5)

If X is an open subset of E then v can be identified with a c^k morphism of X into E, and so, for each c^k morphism f of W into A and each x in W.

$(L_v f)(x) = (Df(x))\,(v(x))$ (9.13.6)

If $L_v f = 0$, for all f in c^k (W,A) then for all x in W, v (x) = 0

Lie Brackets

(9.14) Let X be a c^∞ manifold modelled on a Banach space E over A. Then for every ordered pair (**u**, **v**) of c^∞ vector fields on X there exists a unique vector field, called the Lie Bracket of **u** and **v** denoted by [**u**, **v**], Such that for each open subset W of X and each c^∞ morphism f of W into A.

$L_{[\mathbf{u},\mathbf{v}]} f = L_{\mathbf{u}} (L_{\mathbf{v}} f) - L_{\mathbf{v}} (L_{\mathbf{u}} f)$ and

$$[\mathbf{u},\mathbf{v}]\, f = \mathbf{u}\,(\mathbf{v}\,(f)) - \mathbf{v}\,(\mathbf{u}\,(f)) \qquad (9.14.1)$$

If X is an open subset of **E** then for each c^∞ morphism f of an open subset W of X into A and each x in W.

$$(L_u\,(L_v f) - L_v\,(L_u f))\,(x)$$
$$= (Df(x)\,(D\mathbf{v}(x).\mathbf{u}(x) - D\mathbf{u}(x).\mathbf{v}\,(x))) \quad (9.14.2)$$

and $[\mathbf{u}, \mathbf{v}]\,(x) = D\mathbf{v}\,(x)\,.\,\mathbf{u}\,(x) - D\mathbf{u}\,(x).\,\mathbf{v}\,(x) \qquad (9.14.3)$

Proof : We may suppose that X is an open subset of **E**, W is any open subset of X, f is any c^∞ morphism of W into A and x is any point of W. Then

$$L_u\,(L_v f)\,(x) = D.\,L_v f\,(x).\,\mathbf{u}\,(x)$$

we know $x \overset{Lvf}{\mapsto} (d_x f)\mathbf{v}(x) = Df(x).\mathbf{v}(x)$

We can consider $L_v f$ as the composition of functions h and k given below

$$x \overset{k}{\mapsto} (Df(x), \mathbf{v}(x)) \overset{h}{\mapsto} Df(x).\mathbf{v}(x)$$

Therefore $L_u(L_v f)\,(x) = D.\,L_v f\,(x).\,\mathbf{u}\,(x)$

$= D(h \circ k)(x).\mathbf{u}(x)$

$= Dh\,(k(x)).\,(Dk(x).\mathbf{u}(x))$ (by chain Rule)

$= Dh\,(Df(x), \mathbf{v}(x)).(D^2 f(x).\,\mathbf{u}(x), D\mathbf{v}(x)\,.\,\mathbf{u}(x))$

since h is bilinear function

$= h(Df(x), D\mathbf{v}(x).\mathbf{u}(x)) + h(D^2 f\,(x).\mathbf{u}(x), \mathbf{v}(x)),$

$= Df\,(x)\; D\mathbf{v}(x).\mathbf{u}(x) + D^2 f\,(x).\mathbf{u}(x).\mathbf{v}(x)$

Similarly

$$L_v(L_u f)(x) = D^2 f\,(x).\mathbf{u}\,(x).\,\mathbf{v}\,(x) + Df(x).\,D\mathbf{u}(x).\mathbf{v}(x).$$

Hence

$$L_u(L_v f)(x) - L_v(L_u f)(x) = Df\,(x).\,D\mathbf{v}(x).\mathbf{u}(x) - Df(x).D\mathbf{u}(x).\mathbf{v}(x)$$

or $(L_u(L_v f) - L_v(L_u f))(x) = Df(x).(D\mathbf{v}(x).\mathbf{u}(x) - D\mathbf{u}(x).\mathbf{v}(x))$

let $[\mathbf{u}, \mathbf{v}]$ be the function on X into $\mathbf{E}$ such that for each x in X,

$$[\mathbf{u}, \mathbf{v}](x) = D\mathbf{v}(x).\mathbf{u}(x) - D\mathbf{u}(x).\mathbf{v}(x)$$

Then $[\mathbf{u}, \mathbf{v}]$ is a c^∞ vector field on X and

$$L_{[u, v]} f = L_u (L_v f) - L_v (L_u f)$$

The uniquness and existence of $[\mathbf{u}, \mathbf{v}]$ follows from **(9.13)**.

(9.15) Let $\Gamma^\infty(X)$ be the set of all c^∞ vector fields on a c^∞ manifolds X. Then $\Gamma^\infty(X)$ is an algebra over A, The map $(\mathbf{u}, \mathbf{v}) \mapsto [\mathbf{u}, \mathbf{v}]$ of $\Gamma^\infty(X) \times \Gamma^\infty(X)$ into $\Gamma^\infty(X)$ is alternating and bilinear, and for any three c^∞ vector fields $\mathbf{u}, \mathbf{v}, \mathbf{w}$ on X.

$$[\mathbf{u}, \mathbf{v}] = -[\mathbf{v}, \mathbf{u}] \tag{9.15.1}$$

Jacobi Identity

$$[\mathbf{u}, [\mathbf{v}, \mathbf{w}]] + [\mathbf{v}, [\mathbf{w}, \mathbf{u}]] + [\mathbf{w}, [\mathbf{u}, \mathbf{v}]] = 0 \tag{9.15.2}$$

for any c^∞ vector fields $\mathbf{u}$ and $\mathbf{v}$ on X and any c^∞ map f of X into A

$$[\mathbf{u}, f\mathbf{v}] = f[\mathbf{u}, \mathbf{v}] + (L_u f)\,\mathbf{v} \tag{9.15.3}$$

and

$$[f\mathbf{u}, \mathbf{v}] = f[\mathbf{u}, \mathbf{v}] - (L_v f)(\mathbf{u}) \tag{9.15.4}$$

Let the components of $\mathbf{u}$ and $\mathbf{v}$ with respect to a local coordinate system $(x^1,\ldots,x^n)$ on U be ξ^i and η^i $(i = 1,\ldots,n)$ respectively.

Then by (9.14.3) on U we have for

$$\mathbf{u} = \sum_{i=1}^{n} \xi^i(x)\frac{\partial}{\partial x^i}, \quad \mathbf{v} = \sum_{j=1}^{n} \eta^j(x)\frac{\partial}{\partial x^j}$$

$$[\mathbf{u},\mathbf{v}] = \left[\sum_{i=1}^{n} \xi^i(x)\frac{\partial}{\partial x^i}, \sum_{j=1}^{n} \eta^j(x)\frac{\partial}{\partial x^j}\right]$$

$$= \sum_{i,j}\left[\xi^i(x)\frac{\partial}{\partial x^i}, \eta^j(x)\frac{\partial}{\partial x^j}\right]$$

$$= \sum_{i,j}\left\{\xi^i\eta^j\left[\frac{\partial}{\partial x^i},\frac{\partial}{\partial x^j}\right]+\xi^i\frac{\partial\eta^j}{\partial x^i}\frac{\partial}{\partial x^j}-\eta^j\frac{\partial\xi^i}{\partial x^j}\frac{\partial}{\partial x^i}\right\}$$

Since $\left[\partial/\partial x^i, \partial/\partial x^j\right] = 0$, We have

$$[\mathbf{u},\mathbf{v}] = \sum_{j=1}^{n}\left(\sum_{i=1}^{n}\left(\xi^i\frac{\partial\eta^j}{\partial x^i}-\eta^i\frac{\partial\xi^j}{\partial x^i}\right)\right)\frac{\partial}{\partial x^j} \text{ on U.} \qquad (9.15.5)$$

(9.16) Let m be a c^∞ morphism of a c^∞ manifold X into a c^∞ manifold Y. Let **u** and **v** be any two c^∞ vector fields on X. Let $\mathbf{u}_1 = Tm \circ u \circ m^{-1}$, $\mathbf{v}_1 = Tm \circ v \circ m^{-1}$ be c^∞ vector fields on Y, m-related to **u** and **v** respectively. Then $Tm \circ [\mathbf{u}, \mathbf{v}] \circ m^{-1}$ is a c^∞ vector field on Y and,

$$Tm \circ [\mathbf{u}, \mathbf{v}] \circ m^{-1} = [Tm \circ u \circ m^{-1}, Tm \circ v \circ m^{-1}],$$

i.e.,

$$Tm \circ [\mathbf{u}, \mathbf{v}] = [\mathbf{u}_1, \mathbf{v}_1] \circ m.$$

If m is a c^∞ diffeomorphism of X onto Y then $m_*\mathbf{u}$, $m_*\mathbf{v}$ and $m_*[\mathbf{u}, \mathbf{v}]$ are c^∞ vector fields on Y and $m_*[\mathbf{u}, \mathbf{v}] = [m_*\mathbf{u}, m_*\mathbf{v}]$.

Proof: We can work on charts and so we may suppose that X and Y are open subsets of Banach space E and F over A respectively. Then u and v are c^∞ morphism of X into E.

Let $\mathbf{u}_1 = Tm \circ \mathbf{u} \circ m^{-1}$ and

$\mathbf{v}_1 = Tm \circ \mathbf{v} \circ m^{-1}$, then for all x in X.

$\mathbf{u}_1 \circ m(x) = Tm \circ \mathbf{u}(x) = (Dm(x))(\mathbf{u}(x))$

$\mathbf{v}_1 \circ m(x) = Tm \circ \mathbf{v}(x) = (Dm(x))(\mathbf{v}(x))$

$Tm \circ [\mathbf{u}, \mathbf{v}] \circ m^{-1}(m(x)) = Tm \circ [\mathbf{u}, \mathbf{v}](x) = (Dm(x))([\mathbf{u}, \mathbf{v}](x))$

Hence by (9.14.3) for all x in X.

$[Tm \circ \mathbf{u} \circ m^{-1}, Tm \circ \mathbf{v} \circ m^{-1}](m(x)) = [\mathbf{u}_1, \mathbf{v}_1](m(x))$

$= (D\mathbf{v}_1(m(x)))(\mathbf{u}_1(m(x))) - (D\mathbf{u}_1(m(x)))\,(\mathbf{v}_1(m(x)))$

$= (D\mathbf{v}_1(m(x)))(Dm(x).\mathbf{u}(x)) - (D\mathbf{u}_1(m(x)))(Dm(x).\mathbf{v}(x))$

$= D(\mathbf{v}_1 \circ m)(x).\mathbf{u}(x) - D(\mathbf{u}_1 \circ m)(x).\mathbf{v}(x)$

$= Dm(x).D\mathbf{v}(x).\mathbf{u}(x)+D^2m(x).(\mathbf{u}(x),\mathbf{v}(x))$

$-Dm(x).D\mathbf{u}(x).\mathbf{v}(x)-D^2m(x).(\mathbf{u}(x),\mathbf{v}(x))$

The map $\mathbf{v}_1 \circ m$ appeared above is the map $x \mapsto Dm(x).v(x)$ which can be considered as the composition of the map h and k given below

$$x \overset{k}{\mapsto} (Dm(x),\mathbf{v}(x)) \overset{h}{\mapsto} Dm(x).\mathbf{v}(x)$$

Thus $h \circ k = \mathbf{v}_1 \circ m$

By local Analysis, we have

$D(\mathbf{v}_1 \circ m)(x).\mathbf{u}(x) = D(h \circ k)(x).\mathbf{u}(x)$

$= Dh(k(x))(Dk(x).\mathbf{u}(x))$

$= Dh(Dm(x),\mathbf{v}(x))(D^2m(x).\mathbf{u}(x), D\mathbf{v}(x).\mathbf{u}(x))$

$= h(Dm(x),D\mathbf{v}(x)\mathbf{u}(x)) + h(D^2m(x)).\mathbf{u}(x).\mathbf{v}(x))$

(because h is linear)

$= Dm(x).D\mathbf{v}(x).\mathbf{u}(x) + D^2m(x).(\mathbf{u}(x),v(x))$

similarly,

$D(\mathbf{u}_1 \circ m)(x).\mathbf{v}(x)= Dm(x).D\mathbf{u}(x).\mathbf{v}(x) + D^2m(x).(\mathbf{v}(x),\mathbf{u}(x))$

and so

$[\mathbf{u}_1, \mathbf{v}_1](m(x)) = Dm(x)(D\mathbf{v}(x).\mathbf{u}(x) - D\mathbf{u}(x).\mathbf{v}(x))$

$= Dm(x)[\mathbf{u},\mathbf{v}](x) = Tm \circ [\mathbf{u}, \mathbf{v}](x)$

$= Tm \circ [\mathbf{u}, \mathbf{v}] \circ m^{-1} \circ m(x)$

$= Tm o[\mathbf{u}, \mathbf{v}] \circ m^{-1}(m(x))$, since m is a diffeomorphism.

$[Tm \circ \mathbf{u} \circ m^{-1}, Tm \circ \mathbf{v} \circ m^{-1}] = Tm \circ [\mathbf{u}, \mathbf{v}] \circ m^{-1}$

hence $[m_* \mathbf{u}, m_* \mathbf{v}] = m_* [\mathbf{u}, \mathbf{v}]$.

(9.17) Let $\mathbf{v}$ be a c^∞ map of an open subset X of a Banach space E over A into $\underline{R}$, Let x be any point of X and let a c^∞ map f of $J \times U$ into X be a flow for $\mathbf{v}$ at x, where J is an open interval of A with centre 0 and U is an open neighbourhood of x in X, then

(i) For each c^∞ map g of X into A

$(L_{\mathbf{v}}g)(x) = Dh(0)$

where h is the map $J \ni t \mapsto g(f(t, x)) \in A$

(ii) For each c' map $\mathbf{w}$ of X into $\underline{R}$

$[\mathbf{v}, \mathbf{w}](x) = Dh(0)$

where h is the map $J \ni t \mapsto (D_2 f(-t, x)(\mathbf{w}(f(t, x)))$

Proof: (i) Here the map h can be considered as

$$J \ni t \overset{h}{\mapsto} g \circ f.x(t) \in A$$

So by local analysis, its derivative at 0 is

$$\therefore \quad Dh(0) = D(g \circ f.x)(0) = Dg(f._x(0)) \circ Df.x(0)$$
$$= Dg(f(0, x)) \circ D_1(f(0, x))$$
$$= Dg(f(0, x)) \circ (\mathbf{v}(f(0, x))$$
$$= Dg(x))(\mathbf{v}(x)),$$
$$(f(0,x) = x)$$
$$= L_v g(x)$$

(ii) By local analysis for each t in J. we can determine the derivative Dh(t) as follows, the given function

$$J \ni t \overset{h}{\mapsto} (D_2 f(-t, x))(w(f(t, x))) \in R$$

can be considered as the composition of the maps j and k defined below

$$J \ni t \overset{j}{\mapsto} (D_2 f(-t, x)), (w(f(t, x))) \overset{k}{\mapsto} (D_2 f(-t, x))(w(f(t, x))) \in R$$

Thus we have $k \circ j = h$ so

$$Dh(t) = D(k \circ j)(t) = Dk(j(t)) \circ Dj(t)$$
$$= Dk(D_2 f(-t,x), \mathbf{w}(f(t, x))) \circ (-D_1 D_2 f(-t,x), D(w \circ f._x)(t))$$
$$= k(D_2 f(-t,x), D(\mathbf{w} \circ f.x)(t)) + k(-D_1 D_2 f(-t, x), w(f(t, x)))$$
$$= (D_2 f(-t, x))(D(\mathbf{w} \circ f.x)(t)) - (D_1 D_2 f(-t, x))(\mathbf{w}(f(t, x)))$$
$$= (D_2 f(-t,x))(D\mathbf{w}(f(t, x)) \circ Df._x(t)) - (D_1 D_2 f(-t, x))(\mathbf{w}(f(t, x)))$$
$$= (D_2 f(-t, x))(D\mathbf{w}(f(t, x)))(D_1 f(t, x)) - (D_1 D_2 f(-t, x))(\mathbf{w}(f(t, x)))$$

$$= (D_2f(-t, x))(D\mathbf{w}(f(t, x)))(D_1f(t, x)) - (D\mathbf{v}(f(-t, x)))(\mathbf{w}(f(t, x)))$$

$$\text{at } t = 0, Dh(0) = (D_2f(0, x))(D\mathbf{w}(f(0, x)))(D_1f(0, x)) - (D\mathbf{v}f(0, x))(\mathbf{w}(f(0, x)))$$

Since $f(0,x) = x$, $D_2 f(0, x) = id_{\mathbf{R}}$ and f is a flow for **v** it follows that f is a function $f : J \times U \to X \Rightarrow f.x : J \to X$ is an integral solution for **v**

$$\text{i.e., } \frac{df.x}{dt} = \mathbf{v} \circ f.x\,(t)$$

as X is an open set of E the above condition convert into $Df.x(t) = \mathbf{v} \circ f.x\,(t)$

$$\text{or } D_1f(t,x) = Df.x(t) = \mathbf{v} \circ f.x(t)$$

$$\text{or } D_1 f(0, x) = Df.x(0) = \mathbf{v} \circ f.x(0) = \mathbf{v}(f(0, x)) = \mathbf{v}(x)$$

hence $Dh(0) = D\mathbf{w}(x).\mathbf{v}(x) - D\mathbf{v}(x).\mathbf{w}(x) = [\mathbf{v}, \mathbf{w}](x)$.

(9.18) Let f and g be two one-parameter group of smooth diffeomorpisms of a smooth manifold X, and let u and v be the complete c^∞ vector fields on X generated by f and g. Then [u, v] = 0 iff for all s and t in R.

$$fs. \circ gt. = gt. \circ fs.$$

Proof: We work locally, so withoutloss of generality we may suppose that X is an open subset of a Banach space. Let

$$[u, v] = 0$$

then since g is a one parameter group of smooth diffeomorphisms of X, for all s, t in R

$$gs. \circ gt. = gt. \circ ts. = gs+t$$

hence for all t in R

$$(Tgt.) \circ v \circ gt^{-1} = v$$

and so for all s in R all x in X and all

$v \in \Gamma^\infty(X)$, $D((Tfs.) \circ v \circ fs.^{-1})(x)^2 = 0$

therefore

$$(Tfs.) \circ v \circ fs.^{-1} = v$$

and so for all s and t in R

$$fs. \circ gt. = gt. \circ fs.$$

conversely suppose that for all s, t in R

$$fs. \circ gt. = gt. \circ fs.$$

then for all s in R

$$(Tfs.) \circ v \circ fs.^{-1} = v$$

hence by (9.17)

$$[u, v] = 0$$

10

Differential Forms

Exterior forms arise when we generalized the concepts such as the work of a field along a path and the flux of a fluid through a surface to higher dimensions. Without the knowledge of differential forms Hamiltonian mechanics cannot be understood. The information we need about differential forms involves exterior multiplication, exterior differentiation, integration, and stokes formula.

Exterior Forms

(10.1) *1-forms: Definition*

Let E be an n-dimensional real vector space. We will denote vectors in E by $\mathbf{r}, \mathbf{r}_1, \mathbf{r}_2, \ldots$

A 1-form (or a form of degree 1) is a linear function f: E→R, i.e.,

$$f(a_1\mathbf{r}_1 + a_2\mathbf{r}_2) = a_1 f(\mathbf{r}_1) + a_2 f(\mathbf{r}_2) \quad a_1, a_2 \in \text{R and } \mathbf{r}_1, \mathbf{r}_2 \in \text{E}.$$

We know from linear algebra, that the set of all 1-forms becomes a real vector space for the operations of pointwise sum and scalar multiplication defined by

$$(f_1 + f_2)\mathbf{r} = f_1(\mathbf{r}) + f_2(\mathbf{r}) \text{ and } (af)(\mathbf{r}) = af(\mathbf{r}) \;\forall\, \mathbf{r}$$

The space of 1-forms on E is itself an n-dimensional, and is called the dual space of E denoted by E*

Let $\xi_1,\ldots,\xi_n$ be a linear coordinate system on E, each coordinate $\xi_i : E \to R$ is itself a 1-form, these n 1-forms are linearly independent, therefore, every 1-form f can be expressed in the form $f = a_1\xi_1 + \ldots + a_n\xi_n, \quad (a_i \in \underline{R}, i = 1,\ldots,n)$, the value of f on a vector **r** is equal to

$$f(\mathbf{r}) = a_1\ \xi_1(\mathbf{r}) + \ldots + a_n\ \xi_n(\mathbf{r}).$$

where $\xi_1(\mathbf{r}),\ldots,\xi_n(\mathbf{r})$ are the components of **r** in the chosen coordinate system.

(10.1.1) Example:

If a uniform field **F** is given on an Euclidean space R^3, its work **W** on the displacement **r** is a 1-form acting on **r**.

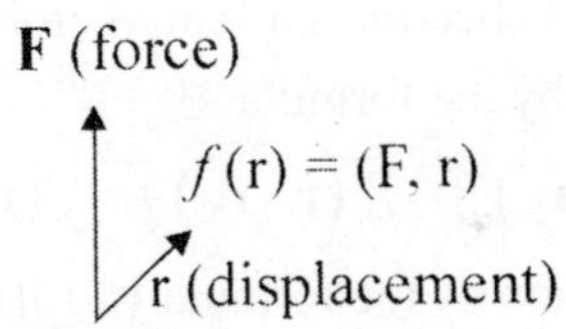

(10.2) ***2-Forms—Definition***

A 2-form (or an exterior form of degree 2) is a function f: E × E → R, which is bilinear and skew symmetric (or antisymmetric), i.e.,

$$f(a_1\mathbf{r}_1 + a_2\mathbf{r}_2, \mathbf{r}_3) = a_1 f(\mathbf{r}_1, \mathbf{r}_3) + a_2 f(\mathbf{r}_2, \mathbf{r}_3)$$

$$f(\mathbf{r}_1, \mathbf{r}_2) = -f(\mathbf{r}_2, \mathbf{r}_1),\ \forall a_1, a_2 \in R, \mathbf{r}_1, \mathbf{r}_2, \mathbf{r}_3 \in E$$

(10.2.1) Example:

Let A $(\mathbf{r}_1, \mathbf{r}_2)$ be the oriented area of the parallelogram constructed on the vectors $\mathbf{r}_1$ and $\mathbf{r}_2$ of the oriented Euclidean plane R^2, i.e.,

$$A(\mathbf{r}_1, \mathbf{r}_2) = \begin{vmatrix} \mathbf{r}_{11} & \mathbf{r}_{12} \\ \mathbf{r}_{21} & \mathbf{r}_{22} \end{vmatrix}$$

Where $\mathbf{r}_1 = \mathbf{r}_{11}\, e_1 + \mathbf{r}_{12}\, e_2$, $\mathbf{r}_2 = \mathbf{r}_{21} + e_1 + \mathbf{r}_{22}\, e_2$

and e_1, e_2 is a basis giving the orientation on R^2.

(10.2.2) Example:

Let v be a uniform velocity vector field for a fluid in three-dimensional oriented Euclidean space, then the flux of the fluid over the area of the parallelogram $\mathbf{r}_1$, $\mathbf{r}_2$ is a bilinear skew symmetric function f of $\mathbf{r}_1$ and $\mathbf{r}_2$, i.e., a 2-form defined by the scalar triple product $f(\mathbf{r}_1, \mathbf{r}_2) = (v, \mathbf{r}_1, \mathbf{r}_2)$.

(10.2.3) Example:

For every 2-form f on E we have $f(\mathbf{r}, \mathbf{r}) = 0 \quad \forall \mathbf{r} \in E$

Solution:

By skew symmetry we have $f(\mathbf{r}, \mathbf{r}) = -f(\mathbf{r}, \mathbf{r}) \Rightarrow 2f(\mathbf{r}, \mathbf{r}) = 0$, the set of all 2-forms on E becomes a real vector space if we define the addition of forms by the formula

$$(f_1 + f_2)(\mathbf{r}_1, \mathbf{r}_2) = f_1(\mathbf{r}_1, \mathbf{r}_2) + f_2(\mathbf{r}_1, \mathbf{r}_2)$$

and scalar multiplication by the formula $(af)(\mathbf{r}_1, \mathbf{r}_2) = af(\mathbf{r}_1, \mathbf{r}_2)$

Let E be a real vector space and $k \geq 0$ an integer, a map f: E×…×E (k factors) to R is k-multilinear, if for each i with $1 \leq i \leq k$ andeach $a \in R$.

$$f(\mathbf{r}_1, \ldots, \mathbf{r}_i + \mathbf{r}_i', \ldots, \mathbf{r}_k) = f(\mathbf{r}_1, \ldots, \mathbf{r}_i, \ldots, \mathbf{r}_k) + f(\mathbf{r}_1, \ldots, \mathbf{r}_i', \ldots, \mathbf{r}_k)$$

and $f(\mathbf{r}_1, \ldots, a\mathbf{r}_i, \ldots, \mathbf{r}_k) = af(\mathbf{r}_1, \ldots, \mathbf{r}_i, \ldots, \mathbf{r}_k)$

for all $(\mathbf{r}_1, \ldots \mathbf{r}_i, \ldots, \mathbf{r}_k)$ and $(\mathbf{r}_1, \ldots, \mathbf{r}_i, \mathbf{r}_i', \ldots, \mathbf{r}_k) \in E$. we denote by $T_k(E; R)$ the set of all such k-multilinear forms which becomes a real vector space with pointwise operations of sum and scalar multiplication

$$(f + g)(\mathbf{r}_1, \ldots, \mathbf{r}_k) = f(\mathbf{r}_1, \ldots, \mathbf{r}_k) + g(\mathbf{r}_1, \ldots, \mathbf{r}_k),$$

$$(af)(\mathbf{r}_1, \ldots, \mathbf{r}_k) = af(\mathbf{r}_1, \ldots, \mathbf{r}_k)$$

Note that T_1 (E ; R) is the dualspace E*.

For k = 0, we write $T_0(E; \underline{R}) = \underline{R}$, the elements of $T_k(E ; \underline{R})$ are called covariant tensor of rank k (or simply k-tensor) on E. If $h : E_1 \to E_2$ is a linear transformation, the pull back map denoted by h*: $T_k(E_2; \underline{R}) \to T_k(E_1; \underline{R})$ and is define for any k as follows,if $f \in T_k(E_2; \underline{R})$, then $h^*(f) \in T_k(E_1; R)$ and so for $\mathbf{r}_1,\ldots,\mathbf{r}_k \in E_1$

$$h^*(f)(\mathbf{r}_1,\ldots,\mathbf{r}_k) = f(h(\mathbf{r}_1),\ldots,h(\mathbf{r}_k))$$

(10.3) ***Definition***:

A mapping $f \in T_k(E; \underline{R})$ is called alternating or skew symmetric if any one of the three equivalent conditions given below are satisfied.

(1) f is zero whenever any two of its arguments are equal, i.e., if $1 \le i, j \le k$ and $i \ne j$, then $f(\mathbf{r}_1,\ldots,\mathbf{r}_i,\ldots,\mathbf{r}_j,\ldots,\mathbf{r}_k) = 0$ whenever $\mathbf{r}_i = \mathbf{r}_j$

(2) f changes sign whenever two of its arguments are interchanged (and the remainng arguments are left fixed). i.e., $f(\mathbf{r}_1,\ldots,\mathbf{r}_j,\ldots,\mathbf{r}_i,\ldots,\mathbf{r}_k) = -f(\mathbf{r}_1,\ldots,\mathbf{r}_i,\ldots,\mathbf{r}_j,\ldots,\mathbf{r}_k)$

(3) If $\sigma \in S_k$ is any permutation of $\underline{k} = \{1,\ldots.k\}$ and $(-1)^\sigma$ is its sign (1 if σ is an even permutation and −1 if σ is an add permutation), then $f(r_{\sigma(1)},\ldots,r_{\sigma(k)}) = (-1)^\sigma f(r_1,\ldots,r_k)$, for K = 1, every linear function f: E → $\underline{R}$ is skew symmetric or alternation. We denote by AT_k (E; $\underline{R}$) the set of k-multilinear skew symmetric maps of $E \times \ldots \times E \xrightarrow{(k-factor)} \underline{R}$, it is a vector subspace of T_k(E; $\underline{R}$) with respect to the pointwise operation of addition and scalar multiplication. The elements of $AT_k(E; \underline{R})$ are called k-forms on E. An elements $f \in T_k(E; \underline{R})$ which takes the same value whenever two of its arguments are interchanged (or, equivalently, whenever its arguments are

permuted) is said to be symmetric and the collection of all such is likewise is a vector subspace of $T_k(E; \underline{R})$. We denote by $ST_k(E; \underline{R})$ the vector subspace of all k-multilinear symmetric mappings.

For any $f \in T_k(E; \underline{R})$ and $g \in T_l(E; \underline{R})$ we define the tensor product

$$f \otimes g \in T_{k+l}(E; \underline{R}) \text{ by}$$

$$(f \otimes g)(\mathbf{r}_1, ..., \mathbf{r}_k, \mathbf{r}_{k+1}, ..., \mathbf{r}_{k+l}) = f(\mathbf{r}_1, ..., \mathbf{r}_k).g(\mathbf{r}_{k+1}, ..., \mathbf{r}_{k+l}) \qquad ...(10.3.1)$$

(10.4) *Lemma*: Let $\{e_1, ..., e_n\}$ be a basis for E and $\{e^1, ..., e^n\}$ its dual basis for T(E; $\underline{R}$) = E* ($e^i(e_j) = \delta^i_j$). Then, for any k, $\{e^{i_1} \otimes, ..., \otimes e^{i_k} : 1 \leq i_1, ..., i_k \leq k\}$ is a basis for $T_k(E; \underline{R})$, moreover, any $f \in T_k(E; \underline{R})$ can be uniquely written as $f = f_{i_1,...,i_k} e^{i_1} \otimes, ..., \otimes e^{i_k}$ where $f_{i_1,...,i_k} = f(e_{i_1}, ..., e_{i_k})$. In particular $\dim T_k(E; \underline{R}) = n^k$ for $n = 2$, the above Lemma gives $\dim T_k(E; R) = 2^k$. In particular, any $f \in T_k(E; \underline{R})$ can be expanded as in Lemma (10.4). However, $e^{i_1} \otimes, ..., \otimes e^{i_k}$ is not skew symmetric, in general, so these do not give a basis for $AT_k(E; R)$. To produce such a basis we intorduced a "skew symmetrized" "alternating" tensor product. Suppose $f \in AT_k(E; \underline{R})$ and $g \in AT_l(E; \underline{R})$. We define their wedge product $f \wedge g$ as follows : If k = 0 or $l = 0$ (or both), we simply let $f \wedge g = fg$, if k > 0 and $l > 0$ we define,

$$(f \wedge g)(\mathbf{r}_1, ..., \mathbf{r}_{k+l}) = \frac{1}{k!l!} \sum_{\sigma} (-1)^{\sigma} (f \otimes g)(\mathbf{r}_{\sigma(11)}, ..., \mathbf{r}_{\sigma(k+l)})$$

$$= \frac{1}{k!l!} \sum_{\sigma} (-1)^{\sigma} f(\mathbf{r}_{\sigma(1)}, ..., \mathbf{r}_{\sigma(k)}) \times g(\mathbf{r}_{\sigma(k+1)}, ..., \mathbf{r}_{\sigma(k+l)})$$

Where the sum is over all permutations $\sigma \in S_{k+l}$ of $\{1, ..., k+l\}$.

(a) For $k = 1$, $l = 1$, i.e., if $f \in AT_1(E; \underline{R})$ and $g \in AT_1(E; \underline{R})$ then $f \wedge g = f \otimes g - g \otimes f$.

(b) For $k = 1$, $l = 2$, i.e., if $f \in T_1(E; \underline{R})$ and $g \in T_2(E; \underline{R})$, then $(f \wedge g)(\mathbf{r}_1, \mathbf{r}_2, \mathbf{r}_3) = f(\mathbf{r}_1)g(\mathbf{r}_2, \mathbf{r}_3) - f(\mathbf{r}_2)g(\mathbf{r}_1, \mathbf{r}_3) + f(\mathbf{r}_3)g(\mathbf{r}_1, \mathbf{r}_2)$

Proof: (a) for $\mathbf{r}_1, \mathbf{r}_2 \in E$

$$(f \wedge g)(\mathbf{r}_1, \mathbf{r}_2) = \frac{1}{1!1!}\sum_{\sigma}(-1)^{\sigma}(f \otimes g)(\mathbf{r}_{\sigma(1)}, \mathbf{r}_{\sigma(2)})$$

$$= (f \otimes g)(\mathbf{r}_1, \mathbf{r}_2) - (f \otimes g)(\mathbf{r}_2, \mathbf{r}_1)$$

$$= (f \otimes g)(\mathbf{r}_1, \mathbf{r}_2) - (g \otimes f)(\mathbf{r}_1, \mathbf{r}_2)$$

$$\Rightarrow f \wedge g = f \otimes g - g \otimes f$$

(b) for $\mathbf{r}_1, \mathbf{r}_2, \mathbf{r}_3 \in E$

$$(f \wedge g)(\mathbf{r}_1, \mathbf{r}_2, \mathbf{r}_3) = \frac{1}{1!2!}\sum_{\sigma}(-1)^{\sigma}(f \otimes g)(x_{\sigma(1)}, x_{\sigma(2)}, x_{\sigma(3)})$$

$$= \frac{1}{2}\begin{bmatrix}(f \otimes g)(\mathbf{r}_1, \mathbf{r}_2, \mathbf{r}_3) + (f \otimes g)(\mathbf{r}_2, \mathbf{r}_3, \mathbf{r}_1) + (f \otimes g)(\mathbf{r}_3, \mathbf{r}_1, \mathbf{r}_2) \\ -(f \otimes g)(\mathbf{r}_1, \mathbf{r}_3, \mathbf{r}_2) - (f \otimes g)(\mathbf{r}_3, \mathbf{r}_2, \mathbf{r}_1) - (f \otimes g)(\mathbf{r}_2, \mathbf{r}_1, \mathbf{r}_3)\end{bmatrix}$$

$$= \frac{2}{2!}\left[(f \otimes g)(\mathbf{r}_1, \mathbf{r}_2, \mathbf{r}_3) - (f \otimes g)(\mathbf{r}_2, \mathbf{r}_1, \mathbf{r}_3) + (f \otimes g)(\mathbf{r}_3, \mathbf{r}_1, \mathbf{r}_2)\right]$$

$$= \left[f(\mathbf{r}_1)g(\mathbf{r}_2, \mathbf{r}_3) - f(\mathbf{r}_2)g(\mathbf{r}_1, \mathbf{r}_3) + f(\mathbf{r}_3)g(\mathbf{r}_1, \mathbf{r}_2)\right]$$

Showing in general case $f \wedge g$ is skew symmetric. We introduce some notation. If $\sigma \in S_k$ is a pe.mutation of $\{1,....,k\}$ and $(\boldsymbol{r}_1,...,\boldsymbol{r}_k)$ is any k-tuple, we will write $\sigma(\mathbf{r}_1,...,\mathbf{r}_k) = (\mathbf{r}_{\sigma(1)},...,\mathbf{r}_{\sigma(k)})$ for the rearranged k-tuple.

For $\sigma, \rho \in S_k$ we have for any k-tuple $(\mathbf{r}_1,...,\mathbf{r}_k)$, $\sigma(\rho(\mathbf{r}_1,...,\mathbf{r}_k)) = (\rho \circ \sigma)(\mathbf{r}_1,...,\mathbf{r}_k)$

Now define, for any $f \in T_k$ (E ; $\underline{R}$), a multilinear function Alt(f) on E × ....×E, (k-factor) → $\underline{R}$ by

$$\text{Alt}(f)(\mathbf{r}_1,...,\mathbf{r}_k) = \frac{1}{k!}\sum_{\sigma \in S_k}(-1)^{\sigma} f(\mathbf{r}_{\sigma(1)},...,\mathbf{r}_{\sigma(k)})$$

Thus, in particular, if $f \in AT_k(E;\underline{R})$ and $g \in AT_l(E;\underline{R})$, then

$$f \wedge g = \frac{(k+l)!}{k!l!}\,\text{Alt}\,(f \otimes g)$$

(10.5) ***Lemma*****:**

Let $k \geq 0$ be an integer, then

(a) $f \in T_k(E;\underline{R}) \Rightarrow \text{Alt}(f) \in AT_k(E;\underline{R})$

(b) $f \in AT_k(E;\underline{R}) \Rightarrow \text{Alt}(f)f$

(c) $f \in T_k(E;\underline{R}) \Rightarrow \text{Alt}(\text{Alt}(f)) = \text{Alt}(f)$

Proof: To prove(a), we prove that Alt (f) is skew symmetric by using definition (10.3) (2) as it is clearly multilinear. First we fix i and j with $1 \leq i,j \leq k$ and show that interchanging the ith and jth arguments change the size of Alt (f). Assume without loss of generality, that $1 \leq i,j \leq k$, and let (i j) be the permutation of {1, ..., k} that permute i and j, but leaves the other fixed. Thus for any k-tuple $(\mathbf{r}_1,...,\mathbf{r}_k)$, $(ij)(\mathbf{r}_1,\mathbf{r}_2,...,\mathbf{r}_k)$ has the ith and jth position element switched while the others are left fixed, moreover, as σ varies over all the permutation in S_k, so does $\sigma' = (ij) \circ \sigma$, but $(-1)^{\sigma'} = -(-1)^{\sigma}$. Thus Alt (f) $(\boldsymbol{r}_1, ...,\boldsymbol{r}_j,...,\boldsymbol{r}_i,...,\boldsymbol{r}_k)$

$$= \frac{1}{k!}\sum_{\sigma}(-1)^{\sigma} f(\sigma.(\mathbf{r}_1,...,\mathbf{r}_j,...,\mathbf{r}_i,...,\mathbf{r}_k))$$

$$= \frac{1}{k!}\sum_{\sigma}(-1)^{\sigma} f(\sigma.((ij)\mathbf{r}_1,...,\mathbf{r}_i,...,\mathbf{r}_j,...,\mathbf{r}_k))$$

$$= \frac{1}{k!}\sum_{\sigma}(-1)^{\sigma} f((ij)\circ\sigma).(\mathbf{r}_1,...,\mathbf{r}_i,...,\mathbf{r}_j,...,\mathbf{r}_k))$$

$$= \frac{1}{k!}\sum_{\sigma'}(-1)^{\sigma'} f(\sigma'.(\mathbf{r}_1,...,\mathbf{r}_i,...,\mathbf{r}_j,...,\mathbf{r}_k))$$

$$= -\frac{1}{k!}\sum_{\sigma'}(-1)^{\sigma'} f(\mathbf{r}_{\sigma'(1)},...,\mathbf{r}_{\sigma'(i)},...,\mathbf{r}_{\sigma'(j)},...,\mathbf{r}_{\sigma'(R)})$$

$$= -\text{Alt}(f)(\mathbf{r}_1,...,\mathbf{r}_i,...,\mathbf{r}_j,...,\mathbf{r}_k)$$

To prove (b) take $f \in AT_k(E;\underline{R})$ and observe that for any $\sigma \in S_k$.

$$f(\sigma.(\mathbf{r}_1,...,\mathbf{r}_k)) = (-1)^\sigma f(\mathbf{r}_1,...,\mathbf{r}_k)$$

Thus $=\frac{1}{k!}\sum_\sigma (-1)^\sigma f(\mathbf{r}_{\sigma(1)},...,\mathbf{r}_{\sigma(k)})$

$$\text{Alt}(f)(\mathbf{r}_1,...,\mathbf{r}_k) = \frac{1}{k!}\sum_\sigma (-1)^\sigma f(\sigma.(\mathbf{r}_1,...,\mathbf{r}_k))$$

$$= \frac{1}{k!}\sum_\sigma (-1)^\sigma (-1)^\sigma f(\mathbf{r}_1,...,\mathbf{r}_k)$$

$$= \frac{1}{k!}k!\,f(\mathbf{r}_1,...,\mathbf{r}_k) = f(\mathbf{r}_1,...,\mathbf{r}_k)$$

(3) obviously follows from (1) & (2).

(10.6) *Example*:

Let $f^1,.....f^k$ be member of $AT_l(E;\underline{R})$, then

$$f^1\wedge...\wedge f^k = \frac{1}{k!}k!\left(f^1\otimes...\otimes f^k\right)$$

$$= \sum_{\sigma\in S_k}(-1)^\sigma . f^{\sigma(1)}\otimes...\otimes f^{\sigma(k)}$$

(10.7) *Theorem*:

Let $\{e_1,...,e_n\}$ be a basis for E and $\{e^1,...,e^n\}$ the corresponding dual basis for $E^* = AT_1(E;\underline{R})$. Then, for each $(k = 1,...,n)$, $\{e^{i_1}\wedge...\wedge e^{i_k} : 1 \le i_1 < .... < i_k \le n\}$ is a basis for $AT_k(E;\underline{R})$ moreover, any $f \in AT_k(E;R)$ can be uniquely expressed as

$$f = \sum_{i_1<...<i_k} f_{i_1...i_k} e^{i_1}\wedge...\wedge e^{i_k} = \frac{1}{k!} f_{i_1...i_k} e^{i_1}\wedge...\wedge e^{i_k}$$

where $f_{i_1...i_k} = f(e_{i_1},...,e_{i_k})$. In particular, dim $AT_k(E;\underline{R}) = \binom{n}{k}$.

Proof: Since $f \in AT_k(E;\underline{R}) \subseteq T_k(E;\underline{R})$.

By lemma (10.4) $f = f_{i_1...i_k}(e^{i_1}\otimes...\otimes e^{i_k})$

where $f_{i_1 \ldots i_k} = f(e_{i_1}, \ldots, e_{i_k})$ and the sum is over all $1 \leq i_1, \ldots, i_k \leq n$, using exercise (10.6) we have

$$f = \frac{1}{k!} f_{i_1, \ldots, i_k} e^{i_1} \wedge \ldots \wedge e^{i_k}.$$

Now, if any term in the last sum has two indices $(i_1, \ldots, i_k)$ equal, the wedge product is zero. Hence, we may sum only those $(i_1, \ldots, i_k)$ that are all dintinct, each such sequence of dintinct indices give rise to k! term in the sum, each term differing from the others only in the order of the indices. By permutating the indices both $f_{i_1, \ldots, i_k}$ and $e^{i_1} \wedge \ldots \wedge e^{i_k}$ changes by the sign of the permutation, so all of these terms are of the same and equal to the unique term

$f_{i_1, \ldots, i_k} e^{i_1} \wedge \ldots \wedge e^{i_k}$ with $i_1 < \ldots < i_k$ conequently

$$f = \sum_{i_1 < \ldots < i_k} f_{i_1, \ldots, i_k} e^{i_1} \wedge \ldots \wedge e^{i_k}$$

In particular, $\{e^{i_1} \wedge \ldots \wedge e^{i_k} \mid 1 \leq i_1 < \ldots < i_k \leq n\}$ spans $AT_k(E; \underline{R})$.

If dim E = n and k > n, then $AT_k(E; \underline{R})$ contain only the zero function. moreover if R = n, the dim $AT_n(E; \underline{R}) = \binom{n}{n} = 1$ so any non-zero element of $AT_n(E; \underline{R})$ spans $AT_n(E, \underline{R})$.

(10.8) *Lemma*

Let $f^1, \ldots, f^k$ be elements of $AT_1(E; \underline{R})$. Then $\{f^1, \ldots, f^k\}$ is linearly independent in $AT_1(E; \underline{R})$ if and only if $f^1 \wedge \ldots \wedge f^k \neq 0$ is in $AT_k(E; \underline{R})$.

Proof: Suppose $\{f^1, \ldots, f^k\}$ is linearly independent. Then there is a basis $\{e_1, \ldots, e_k, \ldots, e_n\}$ for E such that the dual basis $\{e^1, \ldots, e^k, \ldots, e^n\}$ satisfies $e^i = f^i$ for i = 1,..., k. Thus $f^1 \wedge \ldots \wedge f^k$ is an element of a basis for $AT_k(E; \underline{R})$, and so in particular $\neq 0$.

In particular we define exterior product of two 1-forms. Let f_1 and f_2 be two 1-form on E, then, their exterior product is a 2-form on E denoted by $f_1 \wedge f_2$.

The value of the exterior product $f_1 \wedge f_2$ on the pair of vectors $\mathbf{r}_1, \mathbf{r}_2 \in E$ is the oriented area of the image of the parallelogram with sides $f(\mathbf{r}_1)$ and $f(\mathbf{r}_2)$ on the f_1, f_2 plane.

$$(f_1 \wedge f_2)(\mathbf{r}_1, \mathbf{r}_2) = \begin{vmatrix} f_1(\mathbf{r}_1) & f_2(\mathbf{r}_1) \\ f_1(\mathbf{r}_2) & f_2(\mathbf{r}_2) \end{vmatrix}$$

(10.8.1) Exercise:

The mapping

$$(f_1, f_2) \rightarrow f_1 \wedge f_2$$

is bilinear and skew symmetric

$$(f_1 \wedge f_2) = -f_2 \wedge f_1$$

$$(\lambda' f_1' + \lambda'' f_1'') \wedge f_2 = \lambda' f_1' \wedge f_2 + \lambda'' f_1'' \wedge f_2$$

Now choose a system of linear coordinates on E, i.e. let $\mathbf{r}_1, \ldots, \mathbf{r}_n$ be n independent 1-forms. We will call these forms basic.

The exterior product of the two basic forms is the 2-forms $\mathbf{r}_i \wedge \mathbf{r}_j$. By skew-symmetry $\mathbf{r}_i \wedge \mathbf{r}_i = 0$ and $\mathbf{r}_i \wedge \mathbf{r}_j = -\mathbf{r}_j \wedge \mathbf{r}_i$. The geometric meaning of the form $\mathbf{r}_i \wedge \mathbf{r}_j$ is its value on the pair of vectors u, v and is equal to the oriented area of the image of the parallelogram u, v on the coordinate plane $\mathbf{r}_i$, $\mathbf{r}_j$ under the projection parallel to the remaining coordinate directions.

Suppose $f_1, \ldots, f_k$ are k 1-forms we define their exterior product $f_1 \wedge \ldots \wedge f_k$ by setting

$$(f_1 \wedge \ldots \wedge f_k)(\mathbf{r}_1, \ldots, \mathbf{r}_k) = \begin{vmatrix} f_1(\mathbf{r}_1) & \ldots & f_k(\mathbf{r}_1) \\ \vdots & & \vdots \\ f_1(\mathbf{r}_k) & \ldots & f_k(\mathbf{r}_k) \end{vmatrix}$$

In other words, the value of a product of k 1-forms on the parallelopiped $\mathbf{r}_1, \ldots, \mathbf{r}_k$ is equal to the oriented volume of the image of the parallelepiped in the oriented Euclidean coordinate space R^k

under the mapping $\mathbf{r} \to (f_1(\mathbf{r}), \ldots, f_k(\mathbf{r}))$. We call $f_1 \wedge \ldots \wedge f_k$ exterior menomials.

(10.8.2) Example :

Exterior products of k 1-forms is a multilinear skew symmetric mapping $(f_1, \ldots, f_k) \to f_1 \wedge \ldots \wedge f_k$, i.e.,

$$(\lambda' f_1' + \lambda'' f_1'') \wedge f_2 \wedge \ldots \wedge f_k = \lambda' f_1' \wedge f_2 \wedge \ldots \wedge f_k + \lambda'' f_1'' \wedge f_2 \wedge \ldots \wedge f_k$$

and,

$$f_{\sigma(1)} \wedge \ldots \wedge f_{\sigma(k)} = (-1)^{\sigma} f_1 \wedge \ldots \wedge f_k$$

Where $\quad (-1)^{\sigma} = 1$ if σ is an even permutation

$= 1$ if σ is an odd permutation

Now consider the product of two monomials, i.e., a *k*-form or $f_1 \wedge \ldots \wedge f_k = f^k$ and an *l*-form $f_{k+1} \wedge \ldots \wedge f_{k+l} = f^l$ where $f_1, \ldots, f_{k+l}$ are 1-forms. We define their exterior product to be the monomial.

$$\begin{aligned} f^k \wedge f^l &= (f_1 \wedge \ldots \wedge f_k) \wedge (f_{k+1} \wedge \ldots \wedge f_{k+l}) \\ &= f_1 \wedge \ldots \wedge f_k \wedge f_{k+1} \wedge \ldots \wedge f_{k+l} \end{aligned}$$

(10.8.3) Example :

The product of monomials is associative, i.e.,

$$(f^k \wedge f^l) \wedge f^m = f^k \wedge (f^l \wedge f^m)$$

and skew commutative, i.e.,

$$f^k \wedge f^l = (-1)^{kl} f^l \wedge f^k$$

(10.8.4) Example :

Let $p_1, \ldots, p_n, q_1, \ldots, q_n$ are coordinate system on R^{2n}, then the 2-form is given by

$$\sum_{i=1}^{n} p_i \wedge q_i .$$

***(10.8.5) Example*:**

Consider the oriented Euclidean space $\underline{R}^3$. Every vector $v \in \underline{R}^3$ determines a 1-form f_v^1 by setting $f_v^1(\mathbf{r}) = (v, \mathbf{r})$ (scalar product)

$$= v_x \mathbf{r}_x + v_y \mathbf{r}_y + v_z \mathbf{r}_z$$

and a 2-form f_v^2 by setting

$$f_v^2(\mathbf{r}_1, \mathbf{r}_2) = (v, \mathbf{r}_1, \mathbf{r}_2) = \begin{vmatrix} v_x & v_y & v_z \\ \mathbf{r}_{1x} & \mathbf{r}_{1y} & \mathbf{r}_{1z} \\ \mathbf{r}_{2x} & \mathbf{r}_{2y} & \mathbf{r}_{2z} \end{vmatrix}$$

***(10.8.6) Example*:**

Show that the map $v \mapsto f_v^1$ and $v \mapsto f_v^2$ defines isomorphisms of the linear space $\underline{R}^3$ of vectors v with the linear spaces of 1-form on $\underline{R}^3$ and 2-forms on $\underline{R}^3$ respectively. If we choose an orthonormal oriented coordinate system $(\mathbf{r}_{1x}, \mathbf{r}_{1y}, \mathbf{r}_{1z})$ on $\underline{R}^3$, then

$$f_v^1 = v_x \mathbf{r}_{1x} + v_y \mathbf{r}_{1y} + v_z \mathbf{r}_{1z} \quad \text{and}$$
$$f_v^2 = v_x \mathbf{r}_{1y} \wedge \mathbf{r}_{1z} + v_y \mathbf{r}_{1z} \wedge \mathbf{r}_{1x} + v_z \mathbf{r}_{1x} \wedge \mathbf{r}_{1y}$$

(10.9) *Theorem* : Let

$$f, f^1, f^2 \in AT_k(E; \underline{R}), \quad g, g^1, g^2 \in AT_l(E; \underline{R})$$

$$h \in AT_m(E; \underline{R}) \text{ and } a \in \underline{R} \text{ Then}$$

(1) $(f^1 + f^2) \wedge g = f^1 \wedge g + f^2 \wedge g$

(2) $f \wedge (g^1 + g^2) = f \wedge g^1 + f \wedge g^2$

(3) $(af) \wedge g = f \wedge (ag) = a(f \wedge g)$

(4) $g \wedge f = (-1)^{kl} f \wedge g$

(5) $(f \wedge g) \wedge h = f \wedge (g \wedge h) = \dfrac{(k + l + m)!}{k!\ l!\ m!} \mathrm{Alt}(f \otimes g \otimes h)$

(6) If $T : E_1 \to E_2$ is a linear transformation, $f \in AT_k(E_2;R) \wedge$ and

$g \in AT_l(E_2;\underline{R})$, then $T^*(f \wedge g) = (T^* f) \wedge (T^* g)$

Proof : (1), (2) & (3) can be proved easily by using the following results on tensor product,

(1) $(f^1 + f^2) \otimes g = f^1 \otimes g + f^2 \otimes g$

(2) $f \otimes (g^1 + g^2) = f \otimes g^1 + f \otimes g^2$

(3) $(a f) \otimes g = f \otimes (a g) = a(f \otimes g) \qquad (a \in R)$

(4) $f \otimes (g^1 \otimes g^2) = (f \otimes g^1) \otimes g^2$

(5) $T^*(f \otimes g) = (T^* f) \otimes (T^* g)$

We know that, (i) $(f^1 + f^2) \wedge g = \dfrac{(k+l)!}{k!\ l!} \text{Alt}((f^1 + f^2) \otimes g)$

so for $(\mathbf{r}_1, ..., \mathbf{r}_k, \mathbf{r}_{k+1}, ..., \mathbf{r}_{k+l})$,

$$(f^1 + f^2) \wedge g(\mathbf{r}_1, \ldots, \mathbf{r}_k, \mathbf{r}_{k+1}, \ldots, \mathbf{r}_{k+l})$$

$$= \frac{(k+l)!}{k!\ l!} \text{Alt}((f^1 + f^2) \otimes g))(\mathbf{r}_1, \ldots, \mathbf{r}_k, \mathbf{r}_{k+1}, \ldots, \mathbf{r}_{k+1})$$

$$= \frac{(k+l)!}{k!\ l!\ (k+l)!} \sum_{\sigma \in S_{k+l}} (-1)^\sigma ((f^1 + f^2) \otimes g)(\mathbf{r}_{\sigma(1)}, \ldots, \mathbf{r}_{\sigma(k)}, \mathbf{r}_{\sigma(k+1)}, \ldots, \mathbf{r}_{\sigma(k+l)})$$

$$= \frac{1}{k!\ l!} \sum_{\sigma \in S_{k+l}} (-1)^\sigma (f^1 \otimes g + f^2 \otimes g)(\mathbf{r}_{\sigma(1)}, \ldots, \mathbf{r}_{\sigma(k)}, \mathbf{r}_{\sigma(k+1)}, \ldots, \mathbf{r}_{\sigma(k+l)})$$

$$= \frac{1}{k!\ l!} \sum_{\sigma \in S_{k+l}} (-1)^\sigma \left[\begin{matrix} (f^1 \otimes g)(\mathbf{r}_{\sigma(1)}, \ldots, \mathbf{r}_{\sigma(k)}, \mathbf{r}_{\sigma(k+1)}, \ldots, \mathbf{r}_{\sigma(k+l)}) \\ +(f^2 \otimes g)(\mathbf{r}_{\sigma(1)}, \ldots, \mathbf{r}_{\sigma(k)}, \mathbf{r}_{\sigma(k+1)}, \ldots, \mathbf{r}_{\sigma(k+l)}) \end{matrix} \right]$$

$$= ((f^1 \wedge g) + (f^2 \wedge g))(\mathbf{r}_1, \ldots, \mathbf{r}_k, \mathbf{r}_{k+1}, \ldots, \mathbf{r}_{k+l})$$

Hence $(f^1 + f^2) \wedge g = f^1 \wedge g + f^2 \wedge g$, (2) and (3) can be proved similarly as above. For proving (4) we have

$$(f \wedge g)(\mathbf{r}_1, \ldots, \mathbf{r}_k, \mathbf{r}_{k+1}, \ldots, \mathbf{r}_{k+l})$$

$$= \frac{1}{k!\, l!} \sum_{\sigma \in S_{k+l}} (-1)^\sigma f(\mathbf{r}_{\sigma(1)}, \ldots, \mathbf{r}_{\sigma(k)})\, g(\mathbf{r}_{\sigma(k+1)}, \ldots, \mathbf{r}_{\sigma(k+l)})$$

and $(g \wedge f)(r_1, \ldots, r_l, r_{l+k}, \ldots, r_{l+k})$

$$= \frac{1}{l!\, k!} \sum_{\sigma \in S_{k+l}} (-1)^\sigma g(\mathbf{r}_{\sigma(1)}, \ldots, \mathbf{r}_{\sigma(l)})\, f(\mathbf{r}_{\sigma(l+1)}, \ldots, \mathbf{r}_{\sigma(l+k)})$$

$$= \frac{1}{k!\, l!} \sum_{\sigma \in S_{k+l}} (-1)^\sigma f(\mathbf{r}_{\sigma(l+1)}, \ldots, \mathbf{r}_{\sigma(l+k)})\, g(\mathbf{r}_{\sigma(1)}, \ldots, \mathbf{r}_{\sigma(l)})$$

Let ρ be the permutation of $\{1,\ldots,k+l\}$ that sends $\{1,\ldots,l,l+1,\ldots,1+k\}$ in order to $\{l+1,\ldots,l+k,1,\ldots,l\}$. Then ρ is the product of kl transposition (move each of the last k in order over the preceeding l) so $(-1)^\rho = (-1)^{kl}$. Thus for any $\sigma \in S_{k\ l}$, $(-1)^{\sigma\circ\rho} = (-1)^{kl}(-1)^\sigma$. Moreover, as σ varies over S_{k+l}, so does $\sigma' = \sigma \circ \rho$.

Thus, $(g \wedge f)(\mathbf{r}_1, \ldots, \mathbf{r}_{k+1})$

$$= \frac{1}{k!\ l!} \sum_{\sigma \in S_{k+l}} (-1)^\sigma (f \otimes g)(\mathbf{r}_{\sigma(l+1)}, \ldots, \mathbf{r}_{\sigma(l+k)}, \mathbf{r}_{\sigma(1)}, \ldots, \mathbf{r}_{\sigma(l)})$$

$$= \frac{1}{k!\ l!} \sum_{\sigma \in S_{k+l}} (-1)^\sigma (f \otimes g)(\rho \cdot (\mathbf{r}_{\sigma(1)}, \ldots, \mathbf{r}_{\sigma(l)}, \mathbf{r}_{\sigma(l+1)}, \ldots, \mathbf{r}_{\sigma(l+k)}))$$

$$= \frac{1}{k!\ l!} \sum_{\sigma \in S_{k+l}} (-1)^\sigma (f \otimes g)(\rho \cdot (\sigma \cdot (\mathbf{r}_{(1)}, \ldots, \mathbf{r}_{(k+l)})))$$

$$= \frac{1}{k!\ l!} \sum_{\sigma \in S_{k+l}} (-1)^{kl} (-1)^{\sigma\circ\rho} (f \otimes g)((\sigma \circ \rho)(\mathbf{r}_{(1)}, \ldots, \mathbf{r}_{(k+l)}))$$

$$= \frac{(-1)^{kl}}{k!\ l!} \sum_{\sigma \in S_{k+l}} (-1)^{\sigma'} (f \otimes g)(\sigma' \cdot (\mathbf{r}_{(1)}, \ldots, \mathbf{r}_{(k+l)}))$$

$$= (-1)^{kl} (f \wedge g)(\mathbf{r}_{(1)}, \ldots, \mathbf{r}_{(k+l)})$$

(10.10) *Theorem* :

Let $\{e_1,\ldots,e_n\}$ be a basis for E and let $f \in AT_n(E;R)$ If $\mathbf{r}_j = A^i_j e_j$, j = 1,..., n, for any n, vectors in E, then

$$f(\mathbf{r}_1,\ldots,\mathbf{r}_n) = \det(A^i_j)\ f(e_1,\ldots,e_n)$$

Proof : First we see that the ordinary determinent function " det " can be regarded as an element of $AT_n(\underline{R}^n;\ \underline{R})$ as follows: for $\mathbf{r}_1,\ldots,\mathbf{r}_n \in \underline{R}^n$ we write $\mathbf{r}_j = (A^1_j,\ldots,A^n_j)$, (j = 1,...,n) (components relative to the standard basis for R^n) and define

$$\det(\mathbf{r}_1,\ldots,\mathbf{r}_n) = \det\begin{pmatrix} A^1_1 & \cdots & A^1_n \\ \vdots & & \vdots \\ A^n_1 & \cdots & A^n_n \end{pmatrix}$$

Properties of determinant ensure that this is skew symmetric so det $\in AT_n(\underline{R}^n, \underline{R})$. Since det is not identically zero on $\underline{R}^n$,it generates $AT_k(\underline{R}^n;\ \underline{R})$. Now, if $f \in AT_n(E;\ \underline{R})$ we define $f' \in AT_n(\underline{R}^n;\ \underline{R})$ by

$$f^1((A^1_1,\ldots,A^n_1),\ldots,(A^1_n,\ldots,A^n_n)) = f(A^i_1 e_1,\ldots,A^i_n e_i)$$

then $f' = C$ det for some constant C.

Evaluating f' at ((1,0,...,0),...,(0,0,...,1)) given

$$f'((1,0,\ldots,0),\ldots,(0,0,\ldots,1)) = C\ \det(id),$$

$$\text{i.e., } f'(e_1,\ldots,e_n) = C\ .$$

Thus, $f(r_1,\ldots,r_n) = f(A^i_1\ e_i\ ,\ldots,\ A^i_n e_i)$

$$f'((A^1_1,\ldots,A^n_1),\ldots,(A^1_n,\ldots,A^n_n)) = f(e_1,\ldots,e_n)\det(A^i_j)$$

(10.11) *Corollary*

Let $\{e_1,\ldots,e_n\}$ be a basis for E and $\{e^1,\ldots,e^n\}$ be its dual basis for $AT_1(E;\ \underline{R})$ and $(\mathbf{r}_1,\ldots,\mathbf{r}_n)$ vectors in Ewith $\mathbf{r}_j = A^i_j e_i$ (j = 1,...,n) ,Then

$$(e^1\wedge\ldots\wedge e^n)(r_1,\ldots,r_n) = \det(A^i_j)$$

(10.12) *Corollary*:

Let T : E → E be a linear transformation, then

$$T^* : AT_n(E; \underline{R}) \rightarrow AT_n(E ; \underline{R}) \text{ is given by}$$
$$T^* (f) = (\det T) f \text{ for every } f \in AT_n (E; \underline{R})$$

Proof:

If $f = 0$, than $T^* f = 0$ so the result is trivial, suppose that $f \neq 0$, since $\dim AT_n(E;R) = 1$. f generates $AT_n(E; \underline{R})$ so $T^* f = cf$ for some constant c. Let $(e_1,...,e_n)$ be a basis for E and write $Te_j = A^i_j e_i$ $(j = 1,..., n)$, then

$$(T^* f)(e_1,\ldots,e_n) = f(Te_1,\ldots,Te_n)$$
$$= f(A^i_j e_i,\ldots,A^i_n e_i)$$
$$= \det(A^i_j) f(e_1,\ldots,e_n)$$

so $c f(e_1,\ldots,e_n) = \det(A^i_j) f(e_1,\ldots,e_n)$, since

$f \neq 0$, $f(e_1,\ldots,e_n) \neq 0$ so $C = \det(A^i_j)$.

Now we show that any non-zero element of $AT_n(E; R)$ determines a unique orientation for E.

(10.13) *Theorem*:

Let E be an n-dimensional real vector space and f be a non zero element of $AT_n(E; \underline{R})$. Then there is a unique orientation μ for E such that $[e_1,...,e_n] \in \mu$ if and only $f(e_1,...,e_n) > 0$.

Proof:

Since f is non-zero, by theorem **(10.10)** we have $f(e_1,...,e_n)$ is non-zero for every basis $(e_1,\ldots,e_n)$ for E. Thus, the ordered basis for E are divided into two disjoint classes according to whether $f(e_1,...,e_n) > 0$ or $f(e_1,...,e_n) < 0$. Let μ denote the set of all ordered basis for E for which $f(e_1,...,e_n) > 0$. We claim that μ is an orientation for E, i.e., that μ is an equivalence class of the equivalences relation ~ defined on ordered basis as follows : $\{\tilde{e}_1,\ldots,\tilde{e}_n\} \sim \{e_1,\ldots,e_n\}$ iff det $(A^i_j) > 0$ where $\tilde{e}_j = A^i_j e_i (j = 1,\ldots,n)$. But this is clear from theorem **(10.10)** since $f\{\tilde{e}_1,\ldots,\tilde{e}_n\}$ = det $(A^i_j) f(e_1,\ldots,e_n)$ Uniqueness is also clear since E has precisely two orientations and the other one (ie, – μ) consists of the ordered basis for which $f(e_1,\ldots,e_n) < 0$.

Thus if $\{e_1,\ldots,e_n\}$ is an ordered basis for E and $\{e^1,\ldots,e^n\}$ is its dual basis for $AT_1(E; \underline{R})$, then the orientation for E determined by $e^1 \wedge \ldots \wedge e^n$ is precisely the one that contains $\{e_1,\ldots,e_n\}$. Consequently, any non zero n form on E determines unique orientation for E is given. Conversely suppose that, one is given an orientation μ for E. Then every non-zero element of $AT_n(E; \underline{R})$ is either positive on all the elements of μ or negative on all the elements of μ. Thus μ, divides the non-zero element of $AT_n(E; \underline{R})$ into two equivalence classes. In general, however, μ does not singleout a unique element of $AT_n(E; \underline{R})$. However, we show that if E is equipped with the additional structure of an inner product (non degenerate, symmetric, bilinearform, but not necessarily positive definite), then a distinguished element of $AT_n(E; \underline{R})$ is concically determined.

Suppose that E has an orientation μ and an inner product g. If $\{e_1,\ldots,e_n\}$ and $\{\tilde{e}_1,\ldots,\tilde{e}_n\}$ are two basis for E that are orthonormal with respect to g, then the matrix (A^i_j) that relates from ($\tilde{e}_j = A^i_j e_i, (j = 1,\ldots,n)$) satisfies $\det(A^i_j) = \pm 1$ thus for any $f \in AT_n(E; \underline{R})$, $f(\tilde{e}_1,\ldots,\tilde{e}_n) = \pm f(e_1,\ldots,e_n)$.

Now let $\{e_1,\ldots,e_n\}$ be an ordered orthonormal basis in μ. (i.e. an oriented orthonormal basis). Then there exists a non zero element f of AT_n (E ; R) with $f(e_1,\ldots,e_n) = 1$ (namely $e^1 \wedge \ldots \wedge e^n$). But observe that if $(\tilde{e}_1,\ldots,\tilde{e}_n)$ is any other oriented orthonormal basis for E, then $f(\tilde{e}_1,\ldots,\tilde{e}_n) = 1$. Thus f takes any oriented orthonormal basis to 1.

(10.14) *Theorem*:

Let E be as n-dimensional real vector space with an orientation μ and an inner product g. Then ther exists a unique $f \in AT_n(E; \underline{R})$ such that $f(e_1,...,e_n)=1$ whenever $\{e_1,...,e_n\}$ is an oriented orthonormal basis for E.

For example, if $E = \underline{R}^n$ with its standard orientation and (positive definite) innerproduct, then f is just the determinant function $\det \in AT_n(R^n, \underline{R})$. In general f is called the (metric) volume form for E determined by μ and g (recall that in $\underline{R}^3$ $\det |(\mathbf{r}_1, \mathbf{r}_2, \mathbf{r}_3)|$ is the volume of the parallelepiped spanned by $(\mathbf{r}_1, \mathbf{r}_2, \mathbf{r}_3)$). We know that if $\{e_1,...,e_n\}$ is an oriented orthonormal basis for E, then $f = e^1 \wedge ... \wedge e^n$, where $\{e_1,...,e_n\}$ is the dual basis.

We also observe that, If dimE = n and k is any integer with $0 \leq k \leq n$, then (by theorem **(10.7)**,

$$dimAT_{n-k}(E;\ \underline{R}) = \begin{pmatrix} n \\ n-k \end{pmatrix} = \begin{pmatrix} n \\ k \end{pmatrix} = dimAT_k(E;\ \underline{R}).$$

In particular, $AT_{n-k}(E; \underline{R})$ is isomerphic to $AT_k(E; \underline{R})$ for any vector space E. In general, however, there is no 'natural' isomorphism between these two. We show next, however, that if E has an orientation μ and an inner product g, then there is a canonical isomorphism.

$$* : AT_k(E; \underline{R}) \rightarrow AT_{n-k}(E; \underline{R})$$

called the Hodge star operator (the image of $f \in AT_k(E; \underline{R})$ under the isomorphism will be denoted by $*f \in AT_{n-k}(E; \underline{R})$ and called the Hodge dual of f.

(10.14.1) Example:

To convert Laplace Equation in Cartesian coordinates to Polar coordinates:

Solution:

Let $P(x,y)$ and $Q(x,y)$ be two continuous functions in R^2, then

$$*(P\,dx + Q\,dy) = P\,dy - Q\,dx$$

$$*(P\,dx \wedge dy) = P$$

$$*1 = dx \wedge dy.$$

Consider a mapping, $f : R^{+} \times R \to R^2 - \{(0,0)\}$

such that $(r,\theta) \mapsto (r\cos\theta, r\sin\theta)$. ... (10.14.1a)

Now, $$df = \frac{\partial f}{\partial x}dx + \frac{\partial f}{\partial y}dy$$

$$*(df) = -\frac{\partial f}{\partial y}dx + \frac{\partial f}{\partial x}dy$$

$$d*(df) = \left(\frac{\partial^2 f}{\partial x^2} + \frac{\partial^2 f}{\partial y^2}\right)dx \wedge dy$$

$$*d*(df) = \frac{\partial^2 f}{\partial x^2} + \frac{\partial^2 f}{\partial y^2} \qquad \text{... (10.14.1b)}$$

Again, since $$x = r\cos\theta$$

$$y = r\sin\theta$$

Hence, $$dr = \frac{x}{r}dx + \frac{y}{r}dy \quad \text{and}$$

$$d\theta = -\frac{y}{r^2}dx + \frac{x}{r^2}dy$$

$$dr \wedge d\theta = \left(\frac{x}{r}dx + \frac{y}{r}dy\right) \wedge \left(-\frac{y}{r^2}dx + \frac{x}{r^2}dy\right) = \frac{1}{r}dx \wedge dy$$

Thus, $*(dr) = r\,d\theta$,

$$*(d\theta) = -\frac{x}{r^2}dx - \frac{y}{r^2}dy = -\frac{1}{r}dr,$$

$$*(dr \wedge d\theta) = \frac{1}{r}.$$

Now, $*(df) = *(\frac{\partial f}{\partial r}dr + \frac{\partial f}{\partial \theta}d\theta) = (\frac{\partial f}{\partial r}(r\,d\theta) - \frac{1}{r}\frac{\partial f}{\partial \theta}dr)$

$$d*(df) = r\frac{\partial^2 f}{\partial r^2}dr \wedge d\theta + \frac{\partial f}{\partial r}dr \wedge d\theta - \frac{1}{r}\frac{\partial^2 f}{\partial \theta^2}d\theta \wedge dr$$

$$*d*(df) = \frac{\partial^2 f}{\partial r^2} + \frac{1}{r}\frac{\partial f}{\partial r} + \frac{1}{r^2}\frac{\partial^2 f}{\partial \theta^2} \qquad \ldots (10.14.1c)$$

From equation (10.14.1b) and (10.14.1c), we have

$$f^*(\frac{\partial^2 f}{\partial x^2} + \frac{\partial^2 f}{\partial y^2}) = \frac{\partial^2 f}{\partial r^2} + \frac{1}{r}\frac{\partial f}{\partial r} + \frac{1}{r^2}\frac{\partial^2 f}{\partial \theta^2}$$

Where, f is the mapping defined by equation (10.14.1a). Hence,

the Laplace equation $\frac{\partial^2 f}{\partial x^2} + \frac{\partial^2 f}{\partial y^2} = 0$ reduces to

$$\frac{\partial^2 f}{\partial r^2} + \frac{1}{r}\frac{\partial f}{\partial r} + \frac{1}{r^2}\frac{\partial^2 f}{\partial \theta^2} = 0.$$

To construct the isomorphism we must first show that the inner product g on E determines an inner product on each $AT_k(E; \underline{R})$.

Suppose E is an n-dimensional real vector space with an inner product g. We have to show that g induce an inner product (which we also denoted by g) on $AT_k(E; \underline{R})$ for each k = 0,1,...,n. Since $AT_0(E; \underline{R}) = \underline{R}$. We define , for $\alpha, \beta \in AT_0(E; \underline{R})$, $g(\alpha, \beta) = \alpha\beta$. For

$k > 0$ our aim is to define g in such a way that , if $\{e_1,...,e_n\}$is an orthonormal basis for E with dual basis $\{e^1,...,e^n\}$, then $\left\{e^{i_1} \wedge ... \wedge e^{i_k} \mid 1 \le i_1 <...< i_k \le n\right\}$ will be an orthonormal basis for $AT_k(E; \underline{R})$. First we define g in terms of an arbitrary basis for E and show that the definition does not actually depend on the initial choice of the basis and finally prove that the definition has the desired property.

Let $\{e_1,...,e_n\}$ be an arbitrary basis for E (not necessarily orthonormal). Let g_{ij} for i,j = 1,...,n and let $\{g^{ij}\}$ denote the matrix inverse of (g_{ij}). Now, for $\alpha,\beta \in AT_k(E; \underline{R})$, $k \geq 1$, we write

$$\alpha = \frac{1}{k!}\alpha_{i_1...i_k} \cdot e^{i_1} \wedge ... \wedge e^{i_k} \text{ and } \beta = \frac{1}{k!}\beta_{j_1...j_k} \cdot e^{j_1} \wedge ... \wedge e^{j_k}$$

Now define $\alpha^{j_1...j_k} = g^{i_1 \cdot j_1} ... g^{i_k \cdot j_k} \cdot \alpha_{i_1...i_k}$ our definition of g on $AT_k(E; R)$ is now

$$g(\alpha, \beta) = \frac{1}{k!} \alpha^{j_1...j_k} \beta_{j_1...j_k}$$

To show that this definition does not depend on the choice of basis we let $\{\hat{e}_1,...,\hat{e}_n\}$ be another basis with $\hat{e}_j = A^i_j \cdot e_i$, j = 1,...,n. Let (A^j_i) denote the matrix inverse of (A^i_j). Then the dual basis $\{e^1,...,e^n\}$ and $\left\{\hat{e}^1,...,\hat{e}^n\right\}$ are related by $\hat{e}^j = A^j_i \cdot e^i$, j = 1,...,n.

(10.15) Let E and F be two normed vector space over a field A. Let k be any positive integer. Let $T_k(E;F)$ be the normed vector space of all k-multilinear continuous mapping $E^k \to F$; for k = 1 we write T(E; F); for k=0 we define $T_0(E; F)$= F.

Definition: A mapping $f \in T_k(E;F)$ is called alternating if $f(x_1,...,x_k)$ = 0 whenever $x_i=x_{i+1}$ for at least one $i(1 \le i \le k)$. For k = 1 every linear function $E \to F$ is an alternation. Let $AT_k(E;F)$ (respectively $ST_k(E;F)$) denote the vector subspace of $T_k(E;F)$ formed by the k, linear alternating (respectively symmetric) mappings. For k=1, $AT_1(E;F) = T(E;F)$ and for k = 0, we put $AT_0(E;F) = F$.

The vector subspace $AT_k(E;F)$ is closed in $T_k(E;F)$; indeded, suppose $f \in T_k(E;F)$ is the limit of a sequence $f_n \in AT_k(E;F)$: then $\lim_{n\to\infty}\|f_n - f\| = 0$; for $x_1,...,x_k \in E$, the *limit* of $f_n(x_1,...,x_k)$ is $f(x_1,...,x_k)$. Thus if $x_i = x_{i+1}$, it follows that

$$\lim_{n\to\infty} f(x_1,...,x_k) = f(x_1,...,x_k) = 0$$

(10.16) *Permutation group*:

Let S_k denote the group of all permutations on the set $\{1,...,k\}$ of the first k integers > 0. We know that it contains k! elements. A transposition is a permutation $\sigma \in S_n$ such that there exists integers $i,j(i \neq j, 1 \leq i \leq k, 1 \leq j \leq k)$

for which $$\begin{cases} \sigma(i) = j,\ \sigma(j) = i \\ \sigma(p) = p \text{ for } p \neq i, j \end{cases}$$

we see that σ interchange i and j. It is also evident that σ^2 is the identity permutaion. We also know that any permutaion $\sigma \in S_n$ may be written as a product of transposition each of which interchanges two conseccutive integers.

The signature of a permutation σ is denoted by $sig(\sigma) = (-1)^{sig(\sigma)}$ and is equal to $+1$ if σ is an even permujtation, i.e., σ is the products of an even number of transposition, and -1 if σ is an odd permutation, i.e., σ is the product of an odd number of transposition. Thus the mapping $\sigma \mapsto sig(\sigma)$ of the group S_k into the multiplication group of two elements $+1$ and -1 is a homomorphism; and if σ is a transposition, then $sig(\sigma) = -1$.

Let $f \in AT_k(E;F)$ and $g \in AT_n(E;G)$, firstly we introduce a bilinear continuous mapping $\Phi: F \times G \to H$. Which associateds with f and g a mapping $h \in T_{k+n}(E;H)$ given by

$h(x_1,...,x_{k+n}) = \Phi(f(x_1,...,x_k), g(x_{k+1},...,x_{k+n}))$ for $(x_1,...,x_{k+n}) \in E^{k+n}$.

Evidently h is multilinear and continuous.

Let m be another positive integer. We shall denote by $A_{k,m}(E;H)$ the set of all members $h:(x_1,...,x_{k+m})\mapsto f(x_1,...,x_{k+m})$ of $T_{k+m}(E;H)$ such that $(x_1,...,x_k)\mapsto f(x_1,...,x_{k+m})$ and $(x_{k+1},...,x_{k+m})\mapsto f(x_1,...,x_{k+m})$ are alternating.

Evidently $A_{k,m}(E;H)$ is a vector subspace of $T_{k+m}(E;H)$ for each $f\in A_{k,m}(E;H)$. Let $\underset{\sim}{f}$ be the member of $T_{k+m}(E;H)$ such that for each element $(x_1,...,x_{k+m})\in E^{k+m}$

$$\underset{\sim}{f}(x_1,...,x_{k+m})=\sum_{\sigma\in Sh(k,k+m)}Sig(\sigma)f\left(x_{\sigma(!)},...,x_{\sigma(k+m)}\right) \tag{10.16.1}$$

Where $Sh(k,k+m)$ is the set of all permutatuions σ of <u>k+ m</u> such that

$$\sigma(1)<...<\sigma(k) \text{ and } \sigma(k+1)<...<\sigma(k+m) \tag{10.16.2}$$

then $\underset{\sim}{f}$ is alternating.

We must now show that the mapping $\underset{\sim}{f}$ defined by (10.16.1) is an alternation. Given a sequence of vectors $x_1,...,x_{k+m}\in E$ in which two consecutive x's are equal $x_i=x_{i+1}$ it is required to show that $\underset{\sim}{f}(x_1,...,x_{k+m})=0$, i.e., that

$$\sum_{\sigma}Sig(\sigma)f\left(x_{\sigma(1)},...,x_{\sigma(k+m)}\right)=0 \tag{10.16.3}$$

Now divide the permutation σ satisfying (10.16.2) into two cases:

(1) Those σ for which $\sigma^{-1}(i)$ and $\sigma^{-1}(i+1)$ are both integers $\le k$ or $\ge k+1$ .In the first case x_i,x_{i+1} occure amongst the first k places in $f\left(x_{\sigma(1)},...,x_{\sigma(k+m)}\right)$, therefore the corresponding term in (10.16.3) is zero (f being an alternating mappihg of the first k variables). In the second case the corresponding term is also zero for an analogous reason.

(2) The other category is itself divided into two sub-categories,

those σ for which $\sigma^{-1}(i) \leq k$ and $\sigma^{-1}(i+1) \geq k+1$ and those σ for which $\sigma^{-1}(i) \geq k+1$ and $\sigma^{-1}(i+1) \leq k$. Let τ be the transposition which interchanges i and i+1. If σ is in the first subcategory, $\tau\sigma$ is in the second, and vice-versa. We may therefore group in pairs of remaining terms on the left hand side of (10.16.3); for each σ such that $\sigma^{-1}(i) \leq k$ and $\sigma^{-1}(i+1) \geq k+1$ we take

$$\mathrm{Sig}(\sigma)h(x_{\sigma(1)},\ldots,x_{\sigma(k+m)}) - \mathrm{Sig}(\sigma)h(x_{\tau\sigma(1)},\ldots,x_{\tau\sigma(k+m)}) \quad (10.16.4)$$

The sequence $\tau\sigma(1),\ldots,\tau\sigma(k+m)$ is in fact deduced from $\sigma(1),\ldots,\sigma(k+m)$ by interchanging i and i+1, and since $x_i = x_{i+1}$, the value of (10.16.4) is 0.

We shall now give a canonical procedure for associating with every $f \in A_{k,m}(E;H)$ an $\underset{\sim}{f} \in T_{k+m}(E;H)$, The function $a_{k,m}: A_{k,m}(E;H) \to T_{k+m}(E;H)$ which associating each $f \in A_{k,m}(E;H)$ to $\underset{\sim}{f} \in T_{k+m}(E;H)$ is called an alternation.

(10.17) *Definition*:

Let $f \in AT_k(E;F)$ and $g \in AT_m(E;G)$. Then by definition the member $a_{k,m}(f \cdot g)$ of $AT_{k+m}(E;H)$ which assign to each element $(x_1,\ldots,x_{k+m})$ of E^{k+m} the value

$$\left(f \underset{\Phi}{\wedge} g\right)(x_1,\ldots,x_{k+m}) = \left(\underset{\sim}{f \cdot g}\right)(x_1,\ldots,x_{k+m})$$

$$= \sum_{\sigma \in Sh(k,k+m)} \mathrm{Sig}(\sigma)(f \cdot g)\left(x_{\sigma(1)},\ldots,x_{\sigma(k+m)}\right)$$

$$= \sum_{\sigma \in Sh(k,k+m)} f\left(x_{\sigma(1)},\ldots,x_{\sigma(k)}\right) \cdot g\left(x_{\sigma(k+1)},\ldots,x_{\sigma(k+m)}\right)$$

Where the summation over all permutations σ of {1,...,k+m} is called the exterior product of f and g relative to $\Phi : F \times G \to H$ and is denoted by $f \underset{\Phi}{\wedge} g$.

(10.17.1) Example:

When k = m =1, $f \in L(E;F)$ and $g \in L(E;G)$ are linear mapping then $f \hat{}_{\Phi} g \in AT_2(E;H)$ is the bilinear mapping

$$(x_1, x_2) \to \Phi(f(x_1), g(x_2)) - \Phi(f(x_2), g(x_1))$$

When G = R, H = F, the mapping $\Phi : F \times R \to F$ being simply the multiplication of a vector of F by a scalar. In this case we omit Φ in the notion $f \wedge g$, if F = R. The product $f \wedge g$ of two multilinear alternating form is again a multilinear alternating form.

(10.18) *Definition*:

Let $f \in AT_k(E;\underline{R})$ and $g \in AT_m(E;\underline{R})$ then by dedfinition the member $A_{k,m}(f \cdot g)$ of $AT_{k+m}(E;\underline{R})$ which assigns to each element $(x_1, ..., x_{k+m}) \in E^{k+m}$ the value written as

$$(f \wedge g)(x_1, ..., x_{k+m}) = (f \cdot g)(x_1, ..., x_{k+m})$$

$$= \sum_{\sigma \in Sh(k,k+m)} Sig(\sigma)(f \cdot g)\left(x_{\sigma(1)}, ..., x_{\sigma(k+m)}\right)$$

$$= \sum_{\sigma \in Sh(k,k+m)} Sig(\sigma) f\left(x_{\sigma(1)}, ..., x_{\sigma(k)}\right) \cdot g\left(x_{\sigma(k+1)}, ..., x_{\sigma(k+m)}\right)$$

and is called the exterior product of f and g. By multilinear algebra

$$f \wedge g = (-1)^{km} g \wedge f \cdot$$

(10.19) Let U be an open subset of a Banach space E over A. For each positive integer k a function on U into $AT_k(E;F)$ is called an alternating form of degree(or order) k on U with values in Y. A function on U into Y is also called an alternating form of degree(or order) 0 on U with values in Y. A function on U into $AT_k(E;\underline{R})$ is simply called an alternating form of degree(or order) k on U. For each alternating form f of degree k and each alternating form g of degree 0 on U and for each x in U.

$$(f \wedge g)(\mathbf{x}) = (g \wedge f)(\mathbf{x}) = g(\mathbf{x}) \cdot f(\mathbf{x}) = (g \cdot f)(\mathbf{x}).$$

For each integer $k \geq 2$, a function on U into $ST_k(E;F)$ is called a symmetric form of order k on U with values in F. A function on U into $ST_k(E;\underline{R})$ is simply called a symmetric form of order k on U.

(10.20) Let f be a c^m ($m \in \underline{N}$)alternating form of degree k ($k \geq 1$) on an open subset U of a Banach space E over A. For each u in U and each x_0 in E,Df(u).x_0 is a member of $AT_k(E;\underline{R})$ and so the map $f_1(u)(x_0, x_1, \ldots, x_k) \to (Df(u) \cdot x_0)(x_1, \ldots, x_k)$ is a member of $A_{1,k}(E;\underline{R})$ and the composition of the map f_1 and $a_{1,k}$, where $f_1 : U \ni u \mapsto f_1(u) \in A_{1,k}(E;\underline{R})$ and $a_{1,k} : A_{1,k}(E;\underline{R}) \to AT_{k+1}(E;\underline{R})$ is called the exterior derivative of f and is denoted by df, i.e., $df = a_{1,k} \circ f_1$, edvidently for all $u \in U$ $df(u) = a_{1,k}(f_1(u)) = f_1\underset{\sim}{(u)}$ then for all $(x_0, x_1, \ldots, x_k)$ in E^{k+1}

$$(df(u))(x_0, \ldots, x_k) = f_1\underset{\sim}{(u)}(x_0, \ldots, x_k)$$

$$= \sum_{\sigma \in Sh(0,k)} Sig(\sigma) f_1(u)(x_{\sigma(0)}, \ldots, x_{\sigma(k)})$$

Where σ is such that $\sigma(0), \sigma(1) < \sigma(2) < \ldots < \sigma(k)$

$$= \sum_{\sigma \in Sh(0,k)} Sig(\sigma)(Df(u) \cdot x_{\sigma(0)})(x_{\sigma(1)}, \ldots, x_{\sigma(k)})$$

$$= \sum_{0 \leq j \leq k} (-1)^j (Df(u) \cdot x_j)(x_0, \ldots, \hat{x}_j, \ldots, x_k) \qquad (10.20.1)$$

where $(x_0, \ldots, \hat{x}_j, \ldots, x_k) = (x_0, \ldots, x_{j-1}, x_{j+1}, \ldots, x_k)$.

In particular for k = 0 and any u in U and any $x_0 \in X$.

$$(df(u)) \cdot x_0 = Df(u) \cdot x_0$$

evidently df is c^{m-1} on U.

and so if f is of degree 0 then we define df = Df. If g is a c^m alternating form of degree 0 on U then

$$d(f \wedge g) = d(g \cdot f) = d(g \cdot f) + g(df). \qquad (10.20.2)$$

(10.21) Let f be a $c^m(m > 0)$ alternating form of degree k (k > 0) on an open subset U of a Banach space X over A. Let $g_0,\ldots,g_{k-1}$ be k functioin on U into X. Then the function

$$U \ni x \to f(x)\big(g_0(x),\ldots,g_{k-1}(x)\big) \in \underline{R}$$

will be denoted by $f(g_0,\ldots,g_{k-1})$.

Let $g_0,\ldots,g_k$ be any k+1 C^m function on U into X and let x be any point of U. Then for i = 0,...,k by calculus on Banach space

$$\left(D\left(f\left(g_0,\ldots,\hat{g}_i,\ldots,g_k\right)\right)(x)\right)(g_i(x))$$

$$= Df(x)\cdot g_i(x)\cdot\left(g_0(x),\ldots,\hat{g}_i(x),\ldots,g_k(x)\right)$$

$$+\sum_{0\le j\le i}(f(x))\left(g_0(x),\ldots,g_{j-1}(x),Dg_j(x)\cdot g_i(x),g_{j+1}(x),\ldots,\hat{g}_i(x),\ldots,g_k(x)\right)$$

$$+\sum_{0\le j\le i}(f(x))\left(g_0(x),\ldots,\hat{g}_i(x),\ldots,g_{j-1}(x),Dg_j(x)\cdot g_i(x),g_{j+1}(x),\ldots,g_k(x)\right)$$

Hence

$$(df(g_0,\ldots,g_k))(x) = a_{1,k}\left(f_1(g_0,\ldots,g_k)_1(x)\right)$$

$$= a_{1,k}\left(f_1(x)(g_0(x),\ldots,g_k(x))\right)$$

$$= df(x)\cdot(g_0(x),\ldots,g_k(x))$$

$$= \sum_{0\le i\le k}(-1)^i\,(Df(x)\cdot g_i(x))\left(g_0(x),\ldots,\hat{g}_i(x),\ldots,g_k(x)\right)$$

$$= \sum_{0\le i\le k}(-1)^i\left(D\left(f\left(g_0,\ldots,\hat{g}_i,\ldots,g_k\right)\right)(x)\right)(g_i(x))$$

$$-\sum_{0\le i\le k}(-1)^i\sum_{0\le j\le i}(f(x))\left(g_0(x),\ldots,g_{j-1}(x),Dg_j(x)\cdot g_i(x),g_{j+1}(x),\ldots,\hat{g}_i(x),\ldots,g_k(x)\right)$$

$$-\sum_{0\le i\le k}(-1)^i\sum_{i\le j\le k}(f(x))\left(g_0(x),\ldots,\hat{g}_i(x),\ldots,g_{j-1}(x),Dg_j(x)\cdot g_i(x),g_{j+1}(x),\ldots,g_k(x)\right)$$

$$= \sum_{0\le i\le k}(-1)^i\left(D\left(f\left(g_0,\ldots,\hat{g}_i,\ldots,g_k\right)\right)(x)\right)(g_i(x))$$

$$-\sum_{0\le i\le k}(-1)^i\left(\sum_{0\le j\le i}(-1)^j(f(x))\left(Dg_j(x)\cdot g_i(x),g_0(x),\ldots,\hat{g}_j(x),\ldots,\hat{g}_i(x),\ldots,g_k(x)\right)\right)$$

$$-\sum_{0\le i\le k}(-1)^i\left(\sum_{0\le j\le i}(-1)^j(f(x))\left(-Dg_j(x)\cdot g_i(x),g_0(x),\ldots,\hat{g}_i(x),\ldots,\hat{g}_j(x),\ldots,g_k(x)\right)\right)$$

$$= \sum_{0\le i\le k}(-1)^i\left(D\left(f\left(g_0,\ldots,\hat{g}_i,\ldots,g_k\right)\right)(x)\right)(g_i(x))$$

$$+\sum_{0\le i<j\le k}(-1)^{i+j}(f(x))\left(Dg_j(x)\cdot g_i(x)-Dg_i(x)\cdot g_j(x),g_0(x),\ldots,\hat{g}_i(x),\ldots,\hat{g}_j(x),\ldots,g_k(x)\right)$$

$$= \sum_{0\le i\le k}(-1)^i\left(D\left(f\left(g_0,\ldots,\hat{g}_i,\ldots,g_k\right)\right)(x)\right)(g_i(x))$$

$$+\sum_{0\le i\le k}(-1)^{i+j}f\left(\left[g_i,g_j\right],g_0,\ldots,\hat{g}_i,\ldots,\hat{g}_j,\ldots,g_k\right)(x)$$

(10.22) Let W be an open subset of a Banach space X over A. Let f and g be two alternating forms on W of degrees j, $(j > 0)$ and k, $(k > 0)$ respectively. Then the exterior product of f and g, denoted by $f\wedge g$, is the function on W into $AT_{j+k}(X,\underline{R})$ such that for each w in W

$$(f\wedge g)(w) = f(w)\wedge g(w)$$

If f and g are m times continuously differentiable then $f\wedge g$ is also m times continuously differentiable.

(10.23) *Theorem* :

et W be an open subset of a Banach space X over A. Let f and g be C^m $(m > 0)$ differential forms on W of degrees j and k respectively. Then

$$d(f\wedge g) = (df)\wedge g + (-1)^j f\wedge dg\,.$$

Proof: Let w be any point of W and let x be any point of X. Then

$$D(f \wedge g)(w) \cdot x = (Df(w) \cdot x) \wedge g(w) + f(w) \wedge (Dg(w) \cdot x)$$

Here $w \overset{f \wedge g}{\mapsto} f(w) \wedge g(w)$

Let us cosider $f \wedge g$ as the composition of the function.

$$w \overset{k}{\mapsto} (f(w), g(w)) \overset{h}{\mapsto} f(w) \wedge g(w)$$

Then $h \circ k = f \wedge g$

$$\therefore D(f \wedge g)(w) \cdot x = D(h \circ k)(w) \cdot x$$

$$= Dh(f(w), g(w))(Df(w) \cdot x, Dg(w) \cdot x)$$

since h is bilinear

$$= h(Df(w) \cdot x, g(w)) + h(f(w), Dg(w) \cdot x)$$

$$= Df(w) \cdot x \wedge g(w) + f(w) \wedge Dg(w) \cdot x$$

so $D(f \wedge g)(w) \cdot x = (Df(w) \cdot x) \wedge g(w) + (-1)^{jk} (Dg(w) \cdot x) \wedge f(w)$

By multilinear algebra $a_{1,j+k} \circ a_{j,k} = a_{1+j,k} \circ a_{1,j}$

and by definition of exterior product

$$(Df(w) \cdot x) \wedge g(w) = a_{j,k}((Df(w) \cdot x) \cdot g(w))$$

Hence

$$a_{1,j+k}((Df(w) \cdot x) \wedge g(w)) = a_{1,j+k}(a_{j,k}((Df(w) \cdot x) \cdot g(w)))$$

$$= a_{1+j,k}(a_{1,j}((Df(w) \cdot x) \cdot g(w)))$$

$$= a_{1+j,k}(df(w) \cdot g(w)) \cdot x$$

$$= df(w) \wedge g(w) \cdot x$$

Similarly

$$a_{1,j+k}((Dg(w) \cdot x) \wedge f(w)) = dg(w) \wedge f(w) \cdot x$$

By multilinear algebra $dg(w) \wedge f(w) = (-1)^{(k+1)j} f(w) \wedge dg(w)$

Consequently

$$d(f \wedge g)(w) = a_{1,j+k}(D(f \wedge g)(w) \cdot x)$$

$$= a_{1,j+k}\left(\left(Df(w)\cdot x\right)\wedge g(w) + (-1)^{jk}\left(Dg(w)\cdot x\right)\wedge f(w)\right)$$
$$= df(w)\wedge g(w) + (-1)^{jk} dg(w)\wedge f(w)$$
$$= df(w)\wedge g(w) + (-1)^{jk}(-1)^{(k+1)j} f(w)\wedge dg(w)$$
$$= df(w)\wedge g(w) + (-1)^{j} f(w)\wedge dg(w)$$

(10.24) Let U be an open subset of a Banach space X over A. Let f be a C^m $(m > 1)$ alternating form on U of degree k. Then $d(d(f)) = 0$.

Proof: Let u be any point of U, and let x and x′ be any point of X. Then

$$df(u).x = a_{1,k}(Df(u).x)$$

an so in view of the fact that $a_{1,k+1}\circ a_{1,k} = a_{2,k}\circ a_{1,1}$

$$\begin{aligned} d(df)(u).(x, x) &= a_{1,k+1}(D(df)(u).\ x') \\ &= a_{1,k+1}(D(a_{1,k}(Df(u).x)(u).\ x')) \\ &= a_{1,k+1}(a_{1,k}(D(Df(u).x).\ x')) \\ &= a_{2,k}\circ a_{1,1}(D(Df(u).x).x') \\ &= a_{2,k}(a_{1,1}(D(Df(u).x).x')) \\ &= a_{2,k}(D(Df(u).x).x' - D(Df(u).x').x) \\ &= a_{2,k}(0) \\ &= 0 \end{aligned}$$

(10.25) ***Theorem***:

Let W be any open subset of a Banach space X over A. Let $f_1,\ldots,f_n$ be $C^m(m>0)$ alternating forms on W of degrees $k_1,\ldots,k_n$ respectively. Then by (10.24) and induction

$$d(f_1\wedge\ldots\wedge f_n) = df_1\wedge f_2\wedge\ldots\wedge f_n + (-1)^{k_1} f_1\wedge df_2\wedge\ldots\wedge f_n + \ldots + (-1)^{k_1+k_2+\ldots+k_{n-1}} f_1\wedge f_2\wedge\ldots\wedge df_n$$

Proof: Let the result is true for n. Then we shall prove that it is also true for n+1.

$$d(f_1\wedge\ldots\wedge f_n\wedge f_{n+1}) = d(f_1\wedge f_2\wedge\ldots\wedge f_n)\wedge f_{n+1} + \ldots + (-1)^{k_1+\ldots+k_n}(f_1\wedge f_2\wedge\ldots\wedge f_n)\wedge df_{n+1}$$

$$= [(df_1 \wedge f_2 \wedge \ldots \wedge f_n) + (-1)^{k_1} f_1 \wedge df_2 \wedge \ldots \wedge f_n + \ldots + (-1)^{k_1+k_2+\ldots+k_{n-1}} f_1 \wedge f_2 \wedge \ldots \wedge df_n] \wedge f_{n+1} + (-1)^{k_1+\ldots+k_n} (f_1 \wedge f_2 \wedge \ldots \wedge f_n) \wedge df_{n+1}$$

$$= df_1 \wedge f_2 \wedge \ldots \wedge f_n \wedge f_{n+1} + (-1)^{k_1} f_1 \wedge df_2 \wedge \ldots \wedge f_n \wedge f_{n+1} + \ldots + (-1)^{k_1+k_2+\ldots+k_{n-1}} f_1 \wedge f_2 \wedge \ldots \wedge df_n \wedge f_{n+1} + (-1)^{k_1+\ldots+k_n} (f_1 \wedge f_2 \wedge \ldots \wedge f_n) \wedge df_{n+1}$$

(10.26) *Pull Back (Definition)*:

Let X and Y be Banach Space over A, let h be a $C^m(m \in \underline{N}_*)$ function on an open subset U of X into an open subset V of Y, and let f be a tensor field on V of type (0,k) $k \in \underline{N}_*$ with values in a Banach Space Z over A. Then the composition of the function α

$$T_k(Y,Z) \times L(X,Y) \ni (p,q) \mapsto ((x_1,\ldots,x_k) \mapsto p(q(x_1),\ldots,q(x_k))) \in T_k(X,Z)$$

and the function β

$$U \ni x \mapsto ((f \circ h)(x), Dh(x)) \in T_k(Y,Z) \times L(X,Y)$$

is called the inverse image or pullback of f by h, and is denoted by $h^*(f)$ or h^*f. Then for each point x in U and each point $(x_1,\ldots,x_k)$ in X^k

$$(h^*(f)(x))(x_1,\ldots,x_k) = (f(h(x)))(Dh(x)\cdot x_1,\ldots,Dh(x)\cdot x_k) \quad (10.26.1)$$

If h is simply a function on the open subset U of X into the open subset V of Y and if f is a function on V into A, then we define $h^*f = f \circ h$. If f is a k-form on V with values in Z then h^*f is a k-form on U with values in Z. If f is symmetric then h^*f is symmetric.

(10.26.1) Example :

Let f be a mapping from R^2 to R^2, sending polar coordinates of a point (r,θ) to cartesian coordinates (x,y), where

$$x = r\cos\theta \quad \text{and} \quad y = r\sin\theta .$$

then

$$\frac{\partial f}{\partial r}=\frac{\partial f}{\partial x}\frac{\partial x}{\partial r}+\frac{\partial f}{\partial y}\frac{\partial y}{\partial r}$$

or
$$\frac{\partial f}{\partial r}=\frac{\partial f}{\partial x}\cos\theta+\frac{\partial f}{\partial y}\sin\theta$$

or
$$\frac{\partial f}{\partial r}=\frac{x}{r}\frac{\partial f}{\partial x}+\frac{y}{r}\frac{\partial f}{\partial y}$$

or
$$\frac{\partial}{\partial r}=\frac{1}{\sqrt{x^2+y^2}}\left(x\frac{\partial}{\partial x}+y\frac{\partial}{\partial y}\right)$$

Similarly,

$$\frac{\partial}{\partial \theta}=-y\frac{\partial}{\partial x}+x\frac{\partial}{\partial y}$$

Since, $\frac{\partial}{\partial r}$ is not defined at origin, i.e. $\frac{\partial}{\partial r}$ is not a differentiable vector field on R^2. So, we take an open subset U of R^2 as $U=R^2-\{(0,0)\}$ and say that $\frac{\partial}{\partial r}$ is a differentiable vector field on U. Thus $\frac{\partial}{\partial r}$ is a differentiable vector field on U defined by

$$(x,y)\mapsto\left(\frac{x}{\sqrt{x^2+y^2}},\frac{y}{\sqrt{x^2+y^2}}\right).$$

Define a mapping, $\quad f:R^{+}\times R\to R^2-\{(0,0)\}$

by $\quad (r,\theta)\mapsto(r\cos\theta,r\sin\theta)$.

Since, this function goes from polar coordinates (r,θ) to cartesian coordinates (x,y), where $x=r\cos\theta$, $y=r\sin\theta$.

So, $dx = d(r\cos\theta) = \cos\theta\, dr - r\sin\theta\, d\theta$

and $dy = d(r\sin\theta) = \sin\theta\, dr + r\cos\theta\, d\theta$.

Then we define a two-form on cartesian coordinates as $dx \wedge dy$. Hence,

$$\begin{aligned} dx \wedge dy &= (\cos\theta\, dr - r\sin\theta\, d\theta) \wedge (\sin\theta\, dr + r\cos\theta\, d\theta) \\ &= r\cos^2\theta\, dr \wedge d\theta - r\sin^2\theta\, d\theta \wedge dr \\ &= r\ dr \wedge d\theta . \end{aligned}$$

Here, it should be noted that the right hand side of the identity $dx \wedge dy = r\ dr \wedge d\theta$, depends on the way f is defined, but this dependence is not indicated on the left hand side. Hence, we must distinguish $dx \wedge dy$ from $r\ dr \wedge d\theta$. The proper way to express the connection between $dx \wedge dy$ and $r\ dr \wedge d\theta$ is to say :

"$r\ dr \wedge d\theta$ is the Pullback of $dx \wedge dy$ via f".

and we write,

$$f^*(dx \wedge dy) = r\ dr \wedge d\theta .$$

(10.27): Let X and Y be Banach Space over A, let h be a $C^m(m \in \underline{N}_*)$ function on an open subset U of X into an open subset V of Y, and let f and g be alternating form on V of degrees j and $k \geqq 0$ respectively. Then

$$h^*(f \wedge g) = (h^*(f)) \wedge (h^*(g))$$

Proof: Let u be any point of U let $(x_1,\ldots,x_{j+k})$ any point of $X^{j+k}(j,k \geqq 1)$. Then

$$\begin{aligned} ((h^*(f \wedge g))(u))(x_1,\ldots,x_{j+k}) &= ((f \wedge g)(h(u)))(Dh(u)\cdot x_1,\ldots,Dh(u)\cdot x_{j+k}) \\ &= ((f \cdot \underset{\sim}{g})(h(u)))(Dh(u)\cdot x_1,\ldots,Dh(u)\cdot x_{j+k}) \\ &= \sum_{p \in Sh(j,\, j+k)} sgn(p)(f(h(u)))(Dh(u)\cdot x_{p(1)},\ldots,Dh(u)\cdot x_{p(j)}) \\ &\qquad \times (g(h(u)))(Dh(u)\cdot x_{p(j+1)},\ldots,Dh(u)\cdot x_{p(j+k)}) \end{aligned}$$

$$= \sum_{p \in Sh(j, j+k)} \operatorname{sgn}(p)\left(\left(h^*(f)\right)(u)\right)\left(x_{p(1)}, \ldots, x_{p(j)}\right)$$

$$\times\left(\left(h^*(g)\right)(u)\right)\left(x_{p(j+1)}, \ldots, x_{p(j+k)}\right)$$

$$= \left(\left(\left(h^*(f)\right) \wedge \left(h^*(g)\right)\right)(u)\right)\left(x_1, \ldots, x_{j+k}\right)$$

Hence $\quad h^*(f \wedge g) = \left(h^*(f)\right) \wedge \left(h^*(g)\right).$

(10.28) *Theorem*:

Let X and Y be Banach Space over A, let h be a $C^m\left(m \in \underline{\overline{N}}_*\right)$ function on an open subset U of X into an open subset V of Y , and let f be a C^m altenating form of degree $k \geqq 0$ on V. Then $d\left(h^*(f)\right) = h^*(df)$

Proof: first suppose that k = 0. Let u be any point of U. Then for each x in X

$$\left(\left(h^*(df)\right)(u)\right)(x) = \left((df)(h(u))\right)(Dh(u) \cdot x)$$

$$= \left(Df(h(u))\right)(Dh(u) \cdot x)$$

$$= D(f \circ h)(u)$$

$$= D\left(h^*(f)\right)(u) \cdot x$$

$$= d\left(h^*(f)\right)(u) \cdot x$$

Hence $\quad h^*(df) = d\left(h^*(f)\right).$

Now let k be any positive integer . Let u be any point in U, and let $(y_0, y_1, \ldots, y_k)$ be any k+1 points in X. Then by the definition in **(10.26)**, and Local Analysis

$$\left(D\left(h^*(f)\right)(u) \cdot y_0\right)\left(y_1, \ldots, y_k\right) = \left(D(f \circ h)(u) \cdot y_0\right)\left(Dh(u) \cdot y_1 \ldots . Dh(u) \cdot y_k\right)$$

$$+ \sum_{j \in k}\left((f \circ h)(u)\right) \times$$

$$\times\left(Dh(u) \cdot y_1, \ldots, Dh(u) \cdot y_{j-1}, \left(D^2 h(u) \cdot y_0\right) \cdot y_j, Dh(u) \cdot y_{j+1}, \ldots, Dh(u) \cdot y_k\right)$$

Hence for each u in U and each point $(x_0, x_1, \dots, x_k)$ in X^{k+1}

$$\left(d\left(h^*(f)\right)(u)\right)(x_0, x_1, \dots, x_k)$$

$$= \sum_{0 \leq i \leq k} (-1)^i D\left(h^*(f)\right)(u) \cdot x_i (x_0, \dots, \hat{x}_1, \dots, x_k)$$

$$= \sum_{0 \leq i \leq k} (-1)^i \left(D(f \circ h)(u) \cdot x_i\right)\left(Dh(u) \cdot x_0, \dots, \widehat{Dh(u)} \cdot x_1, \dots, Dh(u) \cdot x_k\right)$$

$$+ \sum_{0 \leq i \leq k} (-1)^i \left(\sum_{0 \leq j \leq i} \left((f \circ h)(u)\right) \begin{pmatrix} Dh(u) \cdot x_0, \dots, Dh(u) \cdot x_{j-1}, \left(D^2 h(u) \cdot x_j\right) \cdot x_i, \\ Dh(u) \cdot x_{j+1}, \dots, \widehat{Dh(u)} \cdot x_1, \dots, Dh(u) \cdot x_k \end{pmatrix} \right)$$

$$+ \sum_{0 \leq i \leq k} (-1)^i \left(\sum_{0 \leq j \leq k} \left((f \circ h)(u)\right) \begin{pmatrix} Dh(u) \cdot x_0, \dots, \widehat{Dh(u)} \cdot x_i, \dots, Dh(u) \cdot x_{j-1}, \\ \left(D^2 h(u) \cdot x_j\right) \cdot x_i, Dh(u) \cdot x_{j+1}, \dots, Dh(u) \cdot x_k \end{pmatrix} \right)$$

$$= \sum_{0 \leq i \leq k} (-1)^i \left(D(f \circ h)(u) \cdot x_i\right)\left(Dh(u) \cdot x_0, \dots, \widehat{Dh(u)} \cdot x_1, \dots, Dh(u) \cdot x_k\right)$$

$$- \sum_{0 \leq i \leq k} (-1)^i \left(\sum_{0 \leq j \leq i} (-1)^j \left((f \circ h)(u)\right) \begin{pmatrix} \left(D^2 h(u) \cdot x_i\right) \cdot x_j, Dh(u) \cdot x_0, \dots, \\ \widehat{Dh(u)} \cdot x_j, \dots, \widehat{Dh(u)} \cdot x_i, \dots, Dh(u) \cdot x_k \end{pmatrix} \right)$$

$$- \sum_{0 \leq i \leq k} (-1)^i \left(\sum_{0 \leq j \leq k} (-1)^j \left((f \circ h)(u)\right) \begin{pmatrix} -\left(D^2 h(u) \cdot x_j\right) \cdot x_i, Dh(u) \cdot x_0, \dots, \\ \widehat{Dh(u)} \cdot x_i, \dots, \widehat{Dh(u)} \cdot x_j, \dots, Dh(u) \cdot x_k \end{pmatrix} \right)$$

$$= \sum_{0 \leq i \leq k} (-1)^i \left(D(f \circ h)(u) \cdot x_i\right)\left(Dh(u) \cdot x_0, \dots, \widehat{Dh(u)} \cdot x_1, \dots, Dh(u) \cdot x_k\right)$$

$$- \sum_{0 \leq i < j \leq k} (-1)^{i+j} \left((f \circ h)(u)\right) \begin{pmatrix} \left(D^2 h(u) \cdot x_i\right) \cdot x_j - \left(D^2 h(u) \cdot x_i\right) \cdot x_j, \\ Dh(u) \cdot x_0, \dots, \widehat{Dh(u)} \cdot x_i, \dots \\ \widehat{Dh(u)} \cdot x_j, \dots, Dh(u) \cdot x_k \end{pmatrix}$$

$$= \sum_{0 \leq i \leq k} (-1)^i \left(D(f \circ h)(u) \cdot x_i\right)\left(Dh(u) \cdot x_0, \dots, \widehat{Dh(u)} \cdot x_1, \dots, Dh(u) \cdot x_k\right)$$

$$= \left(df\left(h(u)\right)\right)\left(Dh(u) \cdot x_0, \dots, Dh(u) \cdot x_k\right)$$

$$= \left(\left(h^*(df)\right)(u)\right)(x_0, \dots, x_k)$$

and so

$$d\left(h^* f\right) = h^*(df).$$

(10.29) *Theorem*:

Let X,Y and Z be Banach Space over A. Let h be a differentiable function on an open subset U of X into an open subset V of Y, and let j be a c^1-function on V into an open subset W of Z. Let f be a tensor field on W of type (0,k). Then

$$(j\circ h)^*(f)=\left(h^*\left(j^*(f)\right)\right)$$

Proof: For each point u of U and each point $(x_1,\ldots,x_k)$ in X^k

$$\begin{aligned}&\left(\left((j\circ h)^*(f)\right)(u)\right)(x_1,\ldots,x_k)\\ &=\left(f\left(j\left(h(u)\right)\right)\right)\left((D(j\circ h))(u)\cdot x_1,\ldots,(D(j\circ h))(u)\cdot x_k\right)\\ &=\left(f\left(j\left(h(u)\right)\right)\right)\left(\left(Dj\left(h(u)\right)\right)\cdot\left(Dh(u)\cdot x_1\right),\ldots,\left(Dj\left(h(u)\right)\right)\cdot\left(Dh(u)\cdot x_k\right)\right)\\ &=\left(\left(j^*(f)\right)\left(h(u)\right)\right)\left(\left(Dh(u)\cdot x_1\right),\ldots,\left(Dh(u)\cdot x_k\right)\right)\\ &=\left(\left(h^*\left(j^*(f)\right)\right)(u)\right)(x_1,\ldots,x_k)\end{aligned}$$

Hence

$$(j\circ h)^*(f)=\left(h^*\left(j^*(f)\right)\right)$$

(10.30): Let X be an open subset of A^n. For each i in $\underline{n}$ let pr_i be the projection function on X into A such that for each $(x_j;j\in\underline{n})$ in X, $pr_i(x_j;i\in\underline{n})=x_i$.

Then for each i in $\underline{n}$ dpr_i is a C^∞ form of degree 1 on X, and for each $(x_j;j\in\underline{n})$

$$dpr_i(x_j;j\in\underline{n})=pr_i\,.$$

Let f be an alternating form of degree k(k > 0) on X. Since $(pr_i\,;i\in\underline{n})$ is a basis of the dual $(A^n)^*$ of A^n, by multilinear algebra for each $x=(x_j\,;j\in\underline{n})$ in X

$$f(x)=\sum_{(i_1,\ldots,i_k)\in p_n^k} f(i_1,\ldots,i_k)(x)\,pr_{i_1}\wedge\ldots\wedge pr_{i_k}$$

where p_n^k is the set of all strictly increasing injections of $\underline{k}$ into $\underline{n}$, and for each such injection $(i_1,\ldots,i_k)$ $f(i_1,\ldots,i_k)$ is a function on X into A. Then we often write

$$f = \sum_{(i_1,\ldots,i_k)\in p_n^k} f(i_1,\ldots,i_k)\, pr_{i_1} \wedge \ldots \wedge pr_{i_k}$$

If k = 1 then

$$f = \sum_{i\in \underline{n}} f_i\, dpr_i$$

(10.31): Let X be an open subset of A^n. Let f be a C^1 alternating form of degree 0 on X. Then by local analysis for each x in X

$$df(x) = Df(x) = \sum_{i\in \underline{n}} D_i f(x)\cdot pr_i \;=\sum_{i\in \underline{n}} D_i f(x)\cdot dpr_i(x)$$

Hence

$$df = \sum_{i\in \underline{n}} D_i f\cdot dpr_i$$

Where pr_i is the map of X such that for each $(x_j\,;\, j\in \underline{n})$ in X

$$pr_i((x_j\,;\, j\in \underline{n})) = x_i\,.$$

Next, let $g_1,\ldots,g_k$ by any k C^2 alternating forms of degree 0 on X. Then $dg_1,\ldots,dg_k$ are k C^1 alternating forms of degree 1 on X. Let h be the C^1 alternating form

$$f.dg_1 \wedge \ldots \wedge dg_k$$

of degree k on X. Then by (10.24) and (10.25)

$$\begin{aligned} dh &= df \wedge dg_1 \wedge \ldots \wedge dg_k + fd(dg_1) \wedge dg_2 \wedge \ldots \wedge dg_k \\ &\qquad + \ldots + fdg_1 \wedge \ldots \wedge dg_{k-1} \wedge d(dg_k) \\ &= df \wedge dg_1 \wedge \ldots \wedge dg_k. \end{aligned}$$

Now put k = n, then by multilinear algebra

$$dg_1 \wedge \cdots \wedge dg_n = \mathrm{Det}(D_i g_j\,;\,(i,j)\in \underline{n}\times\underline{n})\, dpr_1 \wedge \cdots \wedge dpr_n.$$

Let h be the n-form

$$h = f.dpr_1 \wedge \ldots \wedge dpr_n\,.$$

and let e be a C^1 map of an open subset X\` of A^n into X. Then by (10.27) and (10.28)

$$e * h = f \circ e.j(e).dpr`_1 \wedge \ldots \wedge dpr`_n,$$

where pr_i is the map of X' such that $pr'_i(x'_j; j \in \underline{n}) = x'_i$, for all $(x'_j;\, j \in \underline{n})$ in X' and all i in $\underline{n}$.

(10.32): Let X be a Banach Space over A. Then a subset W of X is said to be starlike with respect to a point w of it iff for each point x of W the segment [w; x] is contained in W. Every convex subset of X is starlike. Every starlike subset of X is connected.

(10.33): Let X be a Banach Space over A, and let W be any open subset of X. Then for each pair of natural number k and m the set of all C^m differential forms of degree k on W is denoted by $\Omega_k^{(m)}(W)$, which is then a vector space over A. For each natural number m put

$$\Omega^{(m)}(W) = \sum_{k \in \underline{n}} \Omega_k^{(m)}(W)$$

then $\Omega^{(m)}(W)$ is a graded algebra over A with vector multiplication $\wedge$.

(10.34): Let X be a Banach Space over A, let W be an open subset of X such that W is starlike with respect to 0. Let k and m be any positive integers. Then we define a linear map $\Omega_k^{(m)}(W)$ into $\Omega_{k-1}^{(m)}(W)$.

First suppose that k = 1, then for each f in $\Omega_1^{(m)}(W)$ and each x in W put

$$(h(f))(x) = \int_0^1 f(tx) \cdot x dt$$

by local analysis the function

$$W \times [0,1] \ni (x,t) \mapsto f(tx) \cdot x \in A$$

is C^m and so by local analysis h(f) is in $\Omega_0^{(m)}(W)$.

Now let $k > 1$, then for each f in $\Omega_k^{(m)}(W)$, for each x in W and each $(x_1,...,x_{k-1})$ in X^{k-1} put

$$((h(f))(x))(x_1,...,x_{k-1}) = \int_0^1 t^{k-1}(f(tx))(x,x_1,...,x_{k-1})dt \ .$$

Then again by local anlysis for each f in

$$\Omega_k^{(m)}(W), h(f) \text{ is in } \Omega_{k-1}^{(m)}(W).$$

h is evidently linear.

(10.35) *Theorem*:

Let W be an open subset of a Banach Space X over A such that W is starlike with respect to 0 in X. Let k and m be any positive integers. Then for each f in $\Omega_k^{(m)}(W)$

$$h(df) + d(h(f)) = f$$

where h is the linear map of $\Omega_k^{(m)}(W)$ into $\Omega_{k-1}^{(m)}(W)$ defined in **(10.34)**.

Proof: Let f be any member of $\Omega_k^{(m)}(W)$ and let x be any point of W. Then by **(10.20)** for each point $(y_0,...,y_k)$ in X^{k+1}

$$((df)(x))(y_0,...,y_k) = (Df(x).y_0)(y_1,...,y_k)$$
$$+ \sum_{i \in \underline{k}} (-1)^i (Df(x)\cdot y_i)(y_0,...,\hat{y}_i,...,y_k)$$

Hence for each point $(x_1,...,x_k)$ in X^k

$$((h(df))(x))(x_1,...,x_k) = \int_0^1 t^k((df)(tx))(x_1,...,x_k)dt$$

$$= \int_0^1 t^k(Df(tx)\cdot x)(x_1,...,x_k)dt$$

$$+ \sum_{i \in \underline{k}} (-1)^i \int_0^1 t^k(Df(tx)\cdot x_i)(x,x_1,...,\hat{x}_i,...,x_k)dt$$

on the other hand for each point $(y_1,...,y_{k-1})$ in X^{k-1}

$$((h(f))(x))(y_1,...,y_{k-1}) = \int_0^1 t^k(f(tx))(x,y_1,...,y_{k-1})dt$$

and so by differentiation for each point $(y,y_1,...,y_{k-1})$ in X^k

$$(D(h(f))(x).y)(y_1,...,y_{k-1}) = \int_0^1 t^k (Df(tx)\cdot y)(x, y_1,\dots, y_{k-1})dt$$
$$+\int_0^1 t^{k-1}(f(tx))(y, y_1,\dots, y_{k-1})dt$$

Hence by **(10.20)** for each point $(x_1,...,x_k)$ in X^k

$((d(h(f)))(x))(x_1,...,x_k) =$

$$\sum_{i\in\underline{k}}(-1)^{i-1}\big(D(h(f))(x)\cdot x_i\big)(x_1,\dots,\hat{x}_i,\dots,x_k)$$

$$=\sum_{i\in\underline{k}}(-1)^{i-1}\int_0^1 t^k (Df(tx)\cdot x_i)(x, x_1,\dots,\hat{x}_i,\dots,x_k)dt$$
$$+\sum_{i\in\underline{k}}(-1)^{i-1}\int_0^1 t^{k-1}(f(tx))(x_i, x_1,\dots,\hat{x}_i,\dots,x_k)dt$$

$$=\sum_{i\in\underline{k}}(-1)^{i-1}\left(\begin{array}{l}\int_0^1 t^k (Df(tx)\cdot x_i)(x, x_1,\dots,\hat{x}_i,\dots,x_k)dt\\ +k\int_0^1 t^{k-1}(f(tx))(x_1,\dots,x_k)dt\end{array}\right)$$

as f(tx) is an alternating function of $(x_1,...,x_k)$.

Consequently, for each point $(x_1,...,x_k)$ in X^k

$((h(df))(x) + (d(h(f)))(x))(x_1,...,x_k)$

$$=\left(\begin{array}{l}\int_0^1 t^k (Df(tx)\cdot x)(x_1,\dots,x_k)dt\\ +k\int_0^1 t^{k-1}(f(tx))(x_1,\dots,x_k)dt\end{array}\right)$$

$$=\int_0^1\left(t^k (Df(tx)\cdot x)+k\,t^{k-1}(f(tx))\right)(x_1,\dots,x_k)dt$$

$$=\int_0^1\left(\frac{d}{dt}\left(t^k f(tx)\right)\right)(x_1,\dots,x_k)dt$$

$$=\int_0^1\left(\frac{d}{dt}\left(t^k f(tx)\right)dt\right)(x_1,\dots,x_k)$$

$$= (f(x))\ (x_1,...,x_k)$$

so that

$$(h(df))(x) + d(h(f))(x) = f(x).$$

(10.36) *Theorem*:

Let W be an open subset of a Banach Space X over A such that W is starlike with respect to a point in it. Then for each $C^m(m \in \underline{N}_*)$ differential form f on W of degree $k > 0$ such that $df = 0$ there is a C^m differential form g on W of degree k-1 such that $dg = f$.

Proof: Since $df = 0$, by **(10.35)** $f = d(h(f))$.

(10.37): Let X be an open subset of a Banach Space R over A, let v be a $C^m(m \in \underline{N}_*)$ function on X into R, and let w be a C^m function on X into $T_j(R) = T_j(R, A)$, $(j > 0)$. We define a new C^m function $C_v w$ on X into $T_{j-1}(R)$ as follows, First, let $j = 1$, Then for each x in X, $(C_v w)(x) = (w(x))(v(x))$.Next, let $j > 1$, then for each x in X and each element $(v_2,...,v_j)$ in R^{j-1}

$$((C_v w)(x))(v_2,...,v_j) = (w(x))(v(x), v_2,...,v_j)$$

$C_v w$ is a C^m function on X into $T_j(R)$.

Let z be a C^m alternating form of degree k on X, $(k > 0)$. Then by multilinear algebra

$$C_v (w \wedge z) = (C_v w) \wedge z + (-1)^j w \wedge (C_v z).$$

For each form w of degree k $(k \geq 2)$ on X

$$C_v (C_v w) = 0.$$

(10.38): Let X be an open subset of a real Banach Space R, let v be a $C^m(m \in \underline{N}_*)$ function on X into R, and let w be a C^m tensor field of type (0, j) on X, $(j > 0)$.

We define a C^m tensor field of type (0, j) on X, called the Lie derivative of w along v and denoted by $L_v w$, as follows, Let x be any point of X, By local analysis there is an open interval I of $\underline{R}$ containing 0, an open ball W about x contained in X, and a C^∞ function f on $I \times W$ into X such that for each y in W $f(0,y) = y$ and $f_{.y}: t \mapsto f(t,y)$ is an integral solution for v at y ,and such that for t in I $f_{t.}: y \mapsto f(t,y)$ is an isomorphism of an open neighborhood W' of x onto an open neighborhood W' of $f(t, x)$ contained in X, Hence for each i in I Df_t is a linear homeomorphism of R onto itself. Let F be a function on I such that for each t in I

$$F(t) = ((f_t)^* (w \upharpoonright W'))(x) \tag{10.38.1}$$

which is a member of $T_j(R, A)$.

Finally we put

$$(L_v w)(x) = DF(0)$$

Then

$$(L_v w)(x) = \lim_{t \to 0, t \neq 0} \frac{((f_t)(w \upharpoonright w''))(x) - w(x)}{t} \tag{10.38.2}$$

Note that $(L_v w)(x)$ is independent of the choice of I, W and W`.

For each t in I and point $(x_1,...,x_j)$ in R^j

$$(F(t))(x_1,...,x_j) = (w(f_t(x)))(Df_t(x).x_1,...,Df_t(x).x_j)$$
$$= (w(f(t, x)))(D_2f(t, x).x_1,...,D_2f(t, x).x_j)$$

and so

$$(DF(t))(x_1,...,x_j) = (D(w(f(t, x))).D_1f(t, x))(D_2f(t, x).x_1,...,D_2 f(t, x).x_j) + (w(f(t, x)))(D_1D_2f(t, x).x_1, D_2f(t, x).x_2,...,D_2f(t, x).x_j) + ... + (w(f(t, x)))(D_2f(t, x).x_1,...,D_2f(t, x).x_{j-1},D_1 D_2f(t, x).x_j)$$

Using the equations $f(0, x) = x$, $D_1f(0, x) = v(f(t, x))$ and $D_2f(0, x) = id_R$ and the differential equation satisfied by $D_2f(t, x)$ in local analysis

$$((L_v w)(x))(x_1,...,x_j) = (DF(0))(x_1,...,x_j)$$
$$= (Dw(x).v(x))(x_1,...,x_j) + (w(x))(Dv(x).x_1,x_2,...,x_j) + ...$$
$$+ (w(x))(x_1,...,x_{j-1},Dv(x).x_j) \tag{10.38.3}$$

(10.39) ***Theorem*:**

Let X be an open subset of a real Banach Space R, Let v and v` be a C^∞ functions on X into R, let f be a C^∞ function on X into $\underline{R}$ and let w and z be C^∞ differential forms of degree j on X, (j > 0). Then

$$L_v w = C_v(dw) + d(C_v w) \tag{10.39.1}$$
$$L_{fv} w = fL_v w + (df)\wedge C_v w \tag{10.39.2}$$
$$C_{[u;\, v]}(w) = L_u(C_v(w)) - C_v(L_u(w)) \tag{10.39.3}$$
$$L_v(w\wedge z) = (L_v w)\wedge z + w\wedge(L_v z) \tag{10.39.4}$$

Proof: Let x be any point of X, and let $(x_1,\ldots,x_j)$ be any point of R^j $(j>1)$. Then

$$((C_v(dw))(x))(x_1,\ldots,x_j) = ((dw)(x))(v(x),x_1,\ldots,x_j)$$
$$= ((Dw)(x).v(x))(x_1,\ldots,x_j)$$
$$+\sum_{i\in j}(-1)^i(Dw)(x)\cdot x_i\left(v(x),x_1,\ldots,\hat{x}_i,\ldots,x_j\right).$$

on the other hand

$$((d(C_v w))(x))(x_1,\ldots,x_j)$$
$$=\sum_{i\in j}(-1)^{i+1}\left((D(C_v w))(x)\cdot x_i\right)\left(x_1,\ldots,\hat{x}_i,\ldots,x_j\right)$$

But for any point $(y_1,\ldots,y_{j-1})$ in R^{j-1}

$$((C_v w)\,(x))\,(y_1,\ldots,y_{j-1}) = (w(x))(v(x),y_1,\ldots,y_{j-1})$$

and so for any y in R

$$((D(C_v w))\,(x).y)\,(y_1,\ldots,y_{j-1}) = ((Dw)(x).y)(v(x),y_1,\ldots,y_{j-1})$$
$$+(w(x))(Dv(x).y,y_1,\ldots,y_{j-1})$$

Hence

$$((d(C_v w))(x))(x_1,\ldots,x_j)$$
$$=\sum_{i\in j}(-1)^{i+1}\left((Dw)(x)\cdot x_i\right)\left(v(x),x_1,\ldots,\hat{x}_i,\ldots,x_j\right)$$
$$+\sum_{i\in j}(-1)^{i+1}(w(x))\left(Dv(x)\cdot x_i,x_1,\ldots,\hat{x}_i,\ldots,x_j\right)$$

Consequently

$$(C_v(dw))(x) + ((d(C_v w))(x))(x_1,\ldots,x_j)$$
$$= ((Dw)(x).v(x))(x_1,\ldots,x_j) + (w(x)(Dv(x).x_1,x_2,\ldots,x_j)) + \ldots$$
$$+(w(x))(x_1,\ldots,x_{j-1},Dv(x).x_j)$$
$$= ((L_v w)(x))(x_1,\ldots,x_j)$$

Hence

$$L_v w = C_v(dw) + d(C_v w)$$

When $j = 1$ then also

$$L_v w = C_v(dw) + d(C_v w)$$

Thus **(10.39.1)** is proved.

Now by **(10.39.1)**

$$\begin{aligned} L_{fv} w &= C_{fv}(dw) + d(C_{fv} w) \\ &= fC_v(dw) + d(fC_v w) \\ &= fC_v(dw) + (df) \wedge C_v w + fd(C_v w) \\ &= fL_v w + (df) \wedge C_v w \end{aligned}$$

Thus **(10.39.2)** is proved.

By (10.37) and (9.14.3)

$$(C_{[u;\, v]} w)(x)(x_1,\ldots,x_{j-1}) = (w(x))(Dv(x).u(x)-Du(x).v(x).x_1,\ldots,x_{j-1})$$

By (10.38.2)

$$\begin{aligned} (L_u(C_v w)(x))(x_1,\ldots,x_{j-1}) &= (D(C_v w)(x).u(x))(x_1,\ldots,x_{j-1}) \\ &\quad + ((C_v w)(x))(Du(x).x_1,\ldots,x_{j-1}) + \ldots \\ &\quad + ((C_v w)(x))(x_1,\ldots,x_{j-2},Du(x).x_{j-1}) \\ &= (Dw(x).u(x))(v(x),x_1,\ldots,x_{j-1}) \\ &\quad + (w(x))(Dv(x).u(x),x_1,\ldots,x_{j-1}) \\ &\quad + (w(x))(v(x),Du(x).x_1,\ldots,x_{j-1}) + \ldots \\ &\quad + (w(x))(v(x),x_1,\ldots,x_{j-2},Du(x).x_{j-1}) \end{aligned}$$

and

$$\begin{aligned} ((C_v(L_u w))(x))(x_1,\ldots,x_{j-1}) &= ((L_u w)(x))(v(x),x_1,\ldots,x_{j-1}) \\ &= (Dw(x).u(x))(v(x),x_1,\ldots,x_{j-1}) \\ &\quad + (w(x))(Du(x).v(x),x_1,\ldots,x_{j-1}) \\ &\quad + (w(x))(v(x),Du(x).x_1,\ldots,x_{j-1}) + \ldots \\ &\quad + (w(x))(v(x),x_1,\ldots,x_{j-2},Du(x).x_{j-1}) \end{aligned}$$

Hence

$$C_{[u;\, v]} w = L_u(C_v w) - C_v(L_u w)$$

and so **(10.39.3)** is proved.

Finally by **(10.39.1)**

$$
\begin{aligned}
L_v(w \wedge z) &= C_v(d(w \wedge z)) + d(C_v(w \wedge z)) \\
&= C_v((dw) \wedge z + (-1)^j\, w \wedge dz) + d((C_v w) \wedge z) \\
&\qquad\qquad + (-1)^j\, w \wedge C_v z \\
&= (C_v(dw)) \wedge z + (-1)^{j+1}\,(dw) \wedge C_v z \\
&\qquad + (-1)^j\,(C_v w) \wedge dz + (-1)^{2j}\, w \wedge C_v(dz) \\
&\qquad + d((C_v w) \wedge z) + (-1)^{j-1}\,(C_v w) \wedge dz \\
&\qquad + (-1)^j\,(dw) \wedge C_v z + (-1)^{2j}\, w \wedge d(C_v z) \\
&= (C_v(dw)) \wedge z + w \wedge C_v(dz) + (d(C_v w)) \wedge z \\
&\qquad\qquad + w \wedge d(C_v z) \\
&= (L_v w) \wedge z + w \wedge (L_v z)
\end{aligned}
$$

Thus **(10.39.4)** is proved.

(10.40) *Definition*:

Let X be a $C^m(m \in \underline{N}_*)$ manifold and let S be a Banach Space over A. Let $k' \in \underline{N}$, $k' < k$ .Then for positive integer j and j' a $C^{k'}$ section v of the vector bundle

$$\left(T_j^{j'}(\tau_X), T_j^{j'}(T(X), S), X\right)$$

is called a $C^{k'}$ tensor field on X of type (j', j) with values in S. When S = A, the phraze "with values in S" will be deleted . By a $C^{k'}$ tensor field of type (0, 0) on X is meant a $C^{k'}$ morhism of X into A. A $C^{k'}$ tensor field v on X of type (0, j) with values in S is called a differential j-form or simply a j-form on X with values S iff for each x in X v(x) is alternating.

If X is an open subset of a Banach Space R over A then a C^k j-form on X with values in S will be identified with a C^k differential form on X of degree j with values in S.

(10.41) *Thereom*:

Let X be a C^∞ manifold modelled on a Banach Space R over A and Let v be a C^k tensor field of type $(0, j)$ $(j \in \underline{N}_*)$ on X with values in a Banach Space S over A. Then for a list $(u_1,...,u_j)$ of vector fields on an open subset U of X, $v(u_1,...,u_j)$ denotes the map

$$U \ni x \mapsto (v(x))(u_1(x),...,u_j(x)) \in S$$

Let w be another C^k tensor field of type $(0, j)$ on X with values in S, let x be any point of X, such that there exists an open neighbourhood U of x with the property that for all lists $(u_1,...,u_j)$ of vector fields on U.

$$(v(x))(u_1(x),...,u_j(x)) = (w(x))(u_1(x),...,u_j(x))$$

Then $v(x) = w(x)$.

Proof: We may supposed that X is an open subset of R. Assume that $v(x) \neq w(x)$ .Then there exists a list $(x_1,...,x_j)$ of vectors in R such that

$$(v(x))(x_1,...,x_j) \neq (w(x))(x_1,...,x_j)$$

Let u_i be the constant map of U into R such That $u_i(y) = x_i$ for $i = 1,...,j$, and for all y in U. Then

$$(v(x))(u_1(x),...,u_j(x)) \neq (w(x))(u_1(x),...,u_j(x))$$

(10.42): Let v be a $C^{k+1}(k \in \overline{\underline{N}}_*)$ j-form on a C^∞ manifold X modelled on a Banach Space R over A. Then for $j \geq 1$ there exists a unique C^k $(j+1)$-form dv on X called the <u>exterior derivative</u> of v, such that for each open subset U of X and each list $(u_0,u_1,...,u_j)$ of vector fields on U

$$(dv)(u_0,...,u_j) = \sum_{0 \leq m \leq j}(-1)^m L_{u_m}\left(v\left(u_0,...,\hat{u}_m,...,u_j\right)\right)$$

$$+ \sum_{0 \leq m < n \leq j}(-1)^{m+n} v\left([u_m,u_n],u_0,...,\hat{u}_m,...,\hat{u}_n,...,u_j\right)$$

and for $j = 0$

$$(df)(x) = d_x f \text{ for all x in X. (c.f.(5.4))}$$

If X is an open subset of R then for each open subset U of X, Each list $(u_0,...,u_j)$ $(j \geq 1)$ of vector fields on U and each x in U

$((dv)(u_0,...,u_j))(x)$

$$= \sum_{0\le m\le j}(-1)^m\big((Dv(x))(u_m(x))\big)\big(u_0(x),...,\hat{u}_m(x),...,u_j(x)\big)$$

and if v is 0-form then dv = Dv.

Proof :The uniqeness of dv follows form **(10.41):**

The rest follows , when (10.20) is applied to the vector part of v.

(10.43): Let X be a C^∞ manifold,

(1) For each C^1 morphism f of X into A df is the map

$$X \ni x \mapsto d_x f \in (T_x X)$$

(2) For each C^{k+2} ($k \in \underline{N}_*$) j-form v ($j \geq 0$) on X the C^k (j + 2)-form

$$d(d(v)) = 0.$$

(3) For any two C^{k+2} ($k \in \overline{\underline{N}}_*$) j-form v and w on X and any two scalars a and b in A

$$d(av + bw) = adv + bdw$$

(4) For any C^{k+1} i-forms v and C^{k+1} j-form w on X ($i,j \geq 0$) $v \wedge w$ is the C^{k+1} (i+j)-form on X such that for each x in X

$$(v \wedge w)(x) = (v(x)) \wedge (w(x)) \text{ and}$$

$$d(v \wedge w) = (dv) \wedge w + (-1)^j v \wedge (dw).$$

Proof: (10.43) is a straightforward consequences of (10.23) and (10.24).

(10.43(a)): Let h be a C^{k+1} ($k \in \underline{N}_*$) map of a C^{k+1} manifold X modelled on a Banach Space R over A into a C^{k+1} manifold Y modelled on a Banach Space S over A. Let f be a $C^{k'}$ ($k' \in \underline{N}, k' < k$) covariant tensor field on Y of order j ($j \in \underline{N}_*$). Denote by $h^*(f)$ or h^*f the function on X such that for each x in X and each element $(v_1,...,v_j)$ in $(T_x X)^j$.

$$((h^*f)(x))(v_1,...,v_j) = (f(h(x)))(T_x h.v_1,...,T_x h.v_j) \qquad (10.43.1)$$

Then h^*f is a $C^{k'}$ covariant tensor field on X of order j. If g is a C^k map of Y into A then by definition $h^*g = g \circ h$. h^*f is called the <u>pull back</u> of f by h.

If h' is a C^{k+1}($k \in \underline{N}$) map of a C^{k+1} manifold X' modelled on a Banach Space $\underline{R}'$ over A into X them

$$(h \circ h')^*(f) = (h'^*(h^*(f)))$$

Let θ be a chart for Y. Then the j-form $(\theta^{-1})^*(f \upharpoonright Dm\theta)$ on $Im\theta$ is called the local expression of f relative to θ.

By (10.28)

$$d(h^*(f)) = h^*(d(f)) \qquad (10.43.2)$$

If e is a $C^{k'}$ i-form $(i \in \underline{N}_*)$ then

$$h^*(e \wedge f) = (h^*(e)) \wedge (h^*(f))$$

(10.44): Let X be a C^∞ real manifold, Let v be a C^∞ vector field on X, and let w be a C^∞ covariant tensor field of order j,($j \in \underline{N}_*$), on X.

Then there exists a covariant tensor field of order j-1 on X denoted by $C_v w$ or v/w and defined as follows:
for $j \geq 2$ and for x in X each point $(v_1,...,v_{j-1})$ in $(T_x X)^{j-1}$

$$((C_v w)(x))(v_1,...,v_{j-1}) = (w(x))(v(x)v_1,...,v_{j-1}) \qquad (10.44.1)$$

and for j = 1 and for each x in X

$$(C_v w)(x) = (w(x))(v(x)).$$

Let f be the flow for v and let x be any point of X. Then there exists an open ball I about 0 in $\underline{R}$ and an open neighbourhood W of x in X such that $I \times W \subset Dm(f)$ and for each t in I f_t is a diffeomorphism of W onto the neighbourhood $f_t[W]$ of $f_t(x)$.

Let W' be another open neighbourhood of x in X such that $I \times W' \subset Dm(f)$ and for each t in I f_t is a diffeomorphism of W` onto the open neighbourhood $f_t[W]$ of $f_t(x)$.

Then

$$((f_t)^*(w\restriction W))(x) = ((f_t(x)^*(w\restriction W))(x) \tag{10.44.2}$$

and $((f_t)^*(w\restriction W))(x)$ is an element of $T_j(T_xX)$. Then the C^∞ covariant tensor field of order j on X, which assigns to each x in X the element

$$\lim_{t\to 0,\, t\neq 0} \frac{((f_t)(w\restriction W))(x) - w(x)}{t}$$

of $T_j(T_xX)$ is called the <u>derivative</u> of w <u>along</u> v and is denoted by $L_v w$ or vw (by abuse of notation). If z is a C^∞ function on X into <u>R</u> we define

$$L_v z = (dz)(v)$$

w is said to be <u>invariant with respect</u> to v iff $L_v w = 0$.

By (10.38) for each s in I each C^∞ vector field v and each C^∞ j-form w on X and each x in X.

$$((f_s)^*(L_v(w\restriction W)))(x)$$

$$\beta = \lim_{t\to 0,\, t\neq 0} \frac{\left((f_{s+t})^*(w\restriction W)\right)(x) - \left((f_s)^*(w\restriction W)\right)(x)}{t} \tag{10.44.3}$$

By (10.39.2)

$$L_v w = C_v(dw) + d(C_v w) \tag{10.44.4}$$

By (10.44.4) and (10.44.3)

$$d(L_v w) = L_v(dw) \tag{10.44.5}$$

If w is a 1-form

$$C_v w = w(v) \tag{10.44.6}$$

If z is any C^∞ k-form on X, ($k \in \underline{N}_*$) then by (10.39.4) and (10.37)

$$C_v(w\wedge z) = ((C_v w)\wedge z + (-1)^j\, w\wedge(C_v z))$$

$$L_v(w\wedge z) = ((L_v w)\wedge z + w\wedge(L_v z)) \tag{10.44.7}$$

If u is another C^∞ vectorfield on X, then by (10.39.3)

$$C_{[u,v]}(w) = L_u(C_v w) - C_v(L_u w) \tag{10.44.8}$$

If w is closed then

$$L_v w = d(C_v w). \tag{10.44.9}$$

If m is a C^∞ diffeomorphism of X onto a C^∞ manifold Y then

$$(m^{-1})^*(C_v w) = C_{m(v)}((m^{-1})^*(w)) \quad (10.44.10)$$

and

$$(m^{-1})^*(L_v w) = L_{m^*(v)}((m^{-1})^*(w)) \quad (10.44.11)$$

By (10.38) for any j C^∞ vector fields $v_1,...,v_j$ on X

$$(L_v w)(v_1,...,v_j) = L_v(w(v_1,...,v_j)) - \sum_{i\in j} w(v_1,...,[v,v_i],...,v_j) \quad (10.44.12)$$

Let u be another C^∞ vector fields on X, then

(1) If w is invariant with respect to u then by (10.44.8)

$$C_{[u,v]}(w) = L_u(C_v w)$$

(2) If w is invariant with respect to u and v and w is closed then

$$\begin{aligned} C_{[u,v]}(w) &= L_u(C_v w) \\ &= C_u(d(C_v w)) + d(C_u(C_v w)) \\ &= C_u(L_v w) + d(C_u(C_v w)) \\ &= d(C_u(C_v w)) \end{aligned}$$

(3) If w is a 2-form then

$$w(u,v) = - C_u(C_v w) = (C_u w)(v)$$

(4) If w is a closed 2-form then

$$d(w(u,v)) = - d(C_u(C_v w)) = - C_{[u,v]}(w)$$

(5) If w is invariant with respect to u and v and w is a closed 2-form and u′ is any C^∞ vector on X then by (11.44.2)

$$\begin{aligned} L_{u'}(w(u, v)) = (d(w(u, v)))(u') \\ &= - (C_{[u,v]} w)(u') \\ &= - w([u, v], u') \end{aligned}$$

one can define

$$(L_v w)(v_1,...,v_j) = (L_v(w(v_1,...,v_j))) - \sum_{i\in j} w(v_1,...,[v,v_i],...,v_j)$$

and obtain other properties of L_v.

One can define

$$L_v w = C_v(dw) + d(C_v w)$$

whether X is a real or complex C^∞ manifold and obtain other properties of L_v.

(10.45): Let X be a C^∞ manifold modelled on $\underline{R}^n$ and with volume u. Let v be a C^∞ vector field on X. Then ther exists a function, denoted by $\text{Div}_u v$, or simple by Div(v), on X into $\underline{R}$ such that

$$L_v u = (\text{Div}_u v)u \tag{10.45.1}$$

$\text{Div}_u v$ is called the <u>divergence</u> of v <u>with respect to</u> u. Since u is an n-form, du = 0, and so by (10.44)

$$L_v u = C_v(du) + d(C_v u) = d(C_v u)$$

Hence

$$d(C_v u) = (\text{Div}_u v)u \tag{10.45.2}$$

Now suppose that X is an open subset of $\underline{R}^n$ and let

$$u_n = dpr_1 \wedge ... \wedge dpr_n$$

Let x be any point of X. Then

$$u_n(x) = pr_1 \wedge ... \wedge pr_n$$

By (10.44) for each positive integer m < n

$$C_v(u_m \wedge dpr_{m+1}) = (C_v u_m) \wedge dpr_{m+1} + (-1)^m u_m \wedge (C_v(dpr_{m+1}))$$

hence by induction

$$(C_v u_n)(x) = \sum_{i \in \underline{n}} (-1)^{i-1} pr_i(\mathbf{v}(x)) \, pr_1 \wedge ... \wedge \hat{pr_i} \wedge ... \wedge pr_n$$

Then by (10.43)

$$(d(C_v u_n))(x) = \left(\sum_{i \in \underline{n}} D_i(pr_i(\mathbf{v}(x))) \, pr_1 \wedge ... \wedge pr_n \right)$$

Hence

$$\text{Div}_{u_n} v = \sum_{i \in \underline{n}} D_i(pr_i \circ v).$$

(10.46): Let S be a real Banach Space, then a continuous bilinear function f on $S \times S$ into $\underline{R}$ is said to be non-singular (respectively nondegenerate) iff the functions

$$x \mapsto (y \mapsto f(x,y)) \text{ and } x \mapsto (y \mapsto f(y,x))$$

are linear homeomorphisms of S onto its strong dual S^* (respectively continuous linear injections of S into its strong dual S^*). S is said to be selfdual iff there exists a nonsingular continuous bilinear function on $S \times S$ into $\underline{R}$.

If S is finite dimensional then each nondegnerate continuous bilinear function on $S \times S$ into $\underline{R}$ is nonsingular.

(10.47): Let X be a C^k($k \in \underline{N}_*$) manifold modelled on a real Banach Space. A C^{k-1} covariant tensor field f of order two on X is said to be non-singular (respectively nondegenerate) iff for each x in X the map f(x) of $(T_x X) \times (T_x X)$ into $\underline{R}$ is nonsingular (respectively nondegenerate).

(10.48): Let X be a C^∞ manifold modelled on a real Banach Space, and let ω be a C^∞ nonsingular covariant tensor field of order two on X. We denote by ω_x^b the function on TX into T^*X such that for each x in X and each pair of vectors u and v in $T_x X$

$$(\omega_x^b(u))(v) = (\omega(x))(u, v)$$

said to be induced by ω. ω_x^b is evidently a C^∞ isomorphism of TX onto T^*X.

For each C^∞ 1-form u on X the C^∞ vector field $(\omega_x^b)^{-1} \circ u$ on X is said to be induced by u through ω.

For each C^∞ vector field v on X the C^∞ 1-form $\omega_x^b \circ v$ on X is said to be induced by v through ω.

Let g be a C^∞ function on X into $\underline{R}$ Then the vector field on X induced by the 1-form dg on X through ω is called the gradient field induced by g through ω and is denoted by *Grad*(g). Thus

$$Grad(g) = (\omega_x^b)^{-1} \circ (dg)$$

(10.49): Let X be a C^{∞} manifold modelled on a real Banach Space, let ω be a C^{∞} nonsingular 2-form on X, let g and h be C^{∞} 0-form on X, let u be a 1-form on X and let v be a C^{∞} vector field on X, then by using (10.48)

$$\omega_x^b \circ v = C_v\omega \qquad (10.49.1)$$

$$\omega((\omega_x^b)^{-1} \circ u, v) = u(v) \qquad (10.49.2)$$

$$\omega(\textit{Grad}(g), v) = (dg)(v)$$

$$CGrad(g)(\omega) = \omega_x^b \circ Grad(g) = dg$$

$$LGrad(g)(\omega) = CGrad(g)(d\omega)$$

$$L_v g = \omega(Grad(g), v)$$

$$LGrad(g)(g) = 0$$

$$\omega(Grad(g), Grad(h)) = -\,\omega(Grad(h), Grad(g))$$

$$= -\,LGrad(g)(h)$$

$$= -\,CGrad(g)(dh)$$

$$= -\,CGrad(g)(CGrad(g)\omega)$$

(10.50): Let X be a C^{∞} manifold modelled on a real Banach Space R . For each x in X the dual $T^*_x(X)$ of the tangent space T_xX at x is called the <u>co-tangent space</u> at x and an element of T^*_xX is called the <u>co-tangent vector</u> at x .

Put $T^*(X) = \cup(T^*_x(X); x \in X)$ and let τ^*_X be the map of $T^*(x)$ onto X such that for each x in X and each y in $T^*_x(X)$ $\tau^*_X(y) = x$. Then $(\tau^*_X, T^*(X), X)$ is a vector bundle called the <u>co-tangent bundle</u> over X.

For each x in X, each y in $T^*_x(X)$ and each z in $T_y(T^*X)$, $(T(\tau^*_X))(z)$ is in $T_x(X)$ therefore $y((T\tau^*_X)(z))$ is in <u>R</u>, Hence $z \mapsto y((T\tau^*_X)(z))$ is a C^{∞} morphism of $T_y(T^*X)$ into <u>R</u>, and so a member of $T^*_y(T^*X)$. Conseqently $y \mapsto (z \mapsto y((T\tau^*_X)(z)))$ is a C^{∞} morphism of $T^*(X)$ into $T^*(T^*(X))$, that is, a 1-form on $T^*(X)$ called the canonical 1-form on the co-tangent bundle $T^*(X)$ over X.

The exterior derivative of the canonical 1-form on $T^*(X)$ is called the <u>canonical 2-form</u> on the cotangent bundle $T^*(X)$ over X.

Let X be an open subset of R. Then T(X), $T^*(X)$, T(T(X)), $T(T^*(X))$ and $T^*(T^*(X))$ can be identified with $X \times R$, $X \times R^*$, $(X \times R) \times (R \times R)$, $(X \times R^*)$ $(R \times R^*)$ and $(X \times R^*) \times (R \times R^{**})$ respectively.Then for each (x, r^*) in $X \times R^*$

$$\tau^*_X(x, r^*) = x$$

and so

for each $((x, r^*), (r, r^*_1))$ in $(X \times R^*) \times (R \times R^*)$

$$(T(\tau^*_X))((x, r^*), (r, r^*_1)) = (x, r) \tag{10.50.1}$$

Let v be the canonical 1-form on $T^*(x)$. Then

$$(v(x, r^*))((x, r^*), (r, r^*_1)) = r^*(r). \tag{10.50.2}$$

Hence by(10.42)

$$((dv)(x, r^*))(((x, r^*), (r_1, r^*_1)), ((x, r^*), (r_2, r^*_2))) = - r^*_2(r_1) + r^*_1(r_2) \tag{10.50.3}$$

Now let X be C ∞ manifold modelled on $\underline{R}^n(n \in \underline{N}_*)$. Let (U, θ) be a chart for X. Then by (8.6) $(T^*U, (T\theta)^*)$ is a chart for T^*X; in fact $(T\theta)^*$ is a C ∞ diffeomorphism of T^*U onto $\theta[U] \times (\underline{R}^n)^*$. For each y in T^*U and each i in $\underline{n}$ $((T\theta)^*)^i(y) = \theta^i(\tau^*_X(y))$ and $((T\theta)^*)^{(n+i)}(y) = y(T\theta^{-1}(e_i)$, $(e_j; j \in \underline{n})$ being the canonical basis of $\underline{R}^n$.

For each i in $\underline{n}$ let q^i and p_i be the C ∞ function on T^*U into $\underline{R}$ such that for each y in T^*U

$$q^i(y) = ((T\theta)^*)^i (y) \text{ and } p_i (y) = ((T\theta)^*)^{n+i}(y). \tag{10.50.4}$$

Let $(e^j, j \in \underline{n})$ be the basis of $(\underline{R}^n)^*$ dual to $(e_j, j \in \underline{n})$. Let $(f_m; m \in \underline{2n})$ be the basis of $\underline{R}^n \times (\underline{R}^n)^*$ such that for m = 1,...,n $f_m = e_m$ and for m = n + 1,..., n + n $f_m = e^{m-n}$

Then by (5.5) for each y in T^*U

$$((d_y((T\theta)^*))^{-1}(f_m); m \in \underline{2n})$$

is a basis of $T_y(T^*U)$ and $(d_y(T\theta)^*)^m;\ m \in \underline{2n})$ is the basis of $T^*_y(T^*U)$ dual to the above basis of $T_y(T^*U)$, both these bases are called <u>canonical</u>.

Let pr_1 be the projection of $(\underline{R}^n \times (\underline{R}^n)^*)$ onto $\underline{R}^n$ then

$$T(\tau^*_X) = (T\theta^{-1}) \circ pr_1 \circ T((T\theta)^*)$$

By (5.5) for each y in T^*U and each z in $T_y(T^*U)$

$$z = \sum_{m=1}^{n} \left(d_y q^m\right)(z)\left(\left(d_y(T\theta)^*\right)^{-1}(f_m)\right)$$

$$+\sum_{m=1}^{n} \left(d_y p^m\right)(z)\left(\left(d_y(T\theta)^*\right)^{-1}(f_{n+m})\right)$$

And by (5.5)

$$(T(T\theta)^*))(z) = \sum_{m=1}^{n}\left(d_y\left((T\theta)^*\right)^m(z)\right)f_m + \sum_{m=1}^{n}\left(d_y\left((T\theta)^*\right)^{n+m}\right)f_{n+m}$$

hence

$$pr_1((T(T\theta)^*))(z)) = \sum_{m=1}^{n}\left(d_y\left((T\theta)^*\right)^m(z)\right)f_m$$

and so by (10.50.3)

$$y((T\theta^{-1})(pr_1(T(T\theta)^*))(z)) = \sum_{m=1}^{n}\left(d_y\left((T\theta)^*\right)^m(z)\, y(T\theta)^{-1}(f_m)\right)$$

$$= \sum_{m=1}^{n} p^m(y) d_y q^m(z)$$

Thus the canonical 1-form on the cotangent bundle $\overset{*}{TX}$ is given by

$$= \sum_{m=1}^{n} p_m dq^m \qquad (10.50.4)$$

and so the canonical 2-form on $\overset{*}{TX}$ is given by

$$= \sum_{m=1}^{n} dq^m \wedge dp_m \qquad (10.50.5)$$

(10.51) ***Green's Theorem, Gauss's Divergence Theorem and Stoke's Theorem***

In proving the above theorems we required the following:

(A) Differential form in R :

In the study of differential forms on real line **R**, there are two objects of our use :

0-forms : which are just scalar functions of x, i.e. $f(x)$.

1-forms : which are differential of the form $f(x)\,dx$.

(B) Differential form in $\mathbf{R}^2$:

In the study of differential forms on the two dimensional plane, there are three objects of our use :

0-forms: which are just scalar functions of x and y, i.e., $f(x, y)$.

1-forms : which are differentials of the form $f(x, y)\,dx + g(x, y)\,dy$, and

2-forms : which are elements of area, written as $f(x, y)\,dx \wedge dy$.

(C) Differential form in $\mathbf{R}^3$:

Let x^1, x^2 and x^3 be rectangular coordinates on R^3. Then the differential forms which are useful on R^3, are:

0-forms : which are scalar functions of the form $f(x^1, x^2, x^3)$.

1-forms : $f_1\,dx^1 + f_2\,dx^2 + f_3\,dx^3$

2-forms : $f_1\,dx^2 \wedge dx^3 + f_2\,dx^3 \wedge dx^1 + f_3\,dx^1 \wedge dx^2$

3-forms : $f\,dx^1 \wedge dx^2 \wedge dx^3$

where f, f_i's are scalar functions of x^1, x^2 and x^3. Here, it should be noted that no 4-forms or higher order forms are possible in R^3, since these require at least one repeated differential, which gives zero by virtue of (2.1).

Green's Theorem : Let C be a closed C^1 curve in R^2 and D be the interior of C. If $f(x, y)$ and $g(x, y)$ are both C^1 functions defined on D, then

$$\int_C f\,dx + g\,dy = \int_D \left(\frac{\partial g}{\partial y} - \frac{\partial f}{\partial x}\right) dx\,dy \qquad \ldots(10.51.1)$$

Using the differential form the above Green's theorem can be put in compact form as,

$$\int_S d\omega = \int_{\partial s} \omega$$

where ω is the differential 1-form on $S \subset R^2, d\omega$ is the exterior derivative of ω and ∂s represents its boundary (a closed curve in the plane).

Proof : Consider a differential form ω, we may associate with it its exterior derivative $d\omega$. This is obtained by taking, for each term of the differential form ω, the usual differential of the scalar multiplier, and wedging the result into the rest of the term.

Let $\omega = f(x,y)\,dx + g(x,y)\,dy$ is a differential 1-form on $S \subset R^2$, then

$$d\omega = d(f\,dx + g\,dy)$$

$$= df \wedge dx + dg \wedge dy$$

$$= \left(\frac{\partial f}{\partial x} dx + \frac{\partial f}{\partial y} dy\right) \wedge dx + \left(\frac{\partial g}{\partial x} dx + \frac{\partial g}{\partial y} dy\right) \wedge dy$$

$$= \left(\frac{\partial g}{\partial x} - \frac{\partial f}{\partial y}\right) dx \wedge dy \qquad \ldots (10.51.2)$$

Using (10.51.2), Green's theorem (10.51.1), can be written in compact form as,

$$\int_S d\omega = \int_{\partial S} \omega \qquad \ldots (10.51.3)$$

where S, represents a region of the plane and ∂S represents its boundary (a closed curve in the plane).

Stoke's Theorem in R^3: Let $S \subseteq R^3$ be a smooth open surface bounded by a closed curve C. If $\overline{F}(x, y, z)$ be a continuous vector function which has continuous first partial derivatives in $S \subset R^3$,

then

$$\int_C \vec{F}\cdot d\vec{r} = \iint_S (\nabla\times\vec{F})\cdot\hat{n}\,ds \qquad \ldots (10.51.4)$$

where $\hat{n}$ is the outward drawn unit normal vector to $S \subset R^3$.

Using the differential form the above Stoke's theorem can be put in compact form as,

$$\iint_S d\omega = \int_{\partial S} \omega \qquad \ldots (10.51.5)$$

where S, represents a surface in R^3 and ∂S represents its boundary (a closed curve in R^3).

Proof :-In order to obtain the Stoke's theorem from the integral theorem (10.51.3),

let $\omega = f_1\, dx^1 + f_2\, dx^2 + f_3\, dx^3$ (in vector notation it is $\vec{F}\cdot d\vec{r}$)

... (10.51.6)

be a differential 1-form in R^3, where f_i's are functions of x^1, x^2, x^3, then

$$d\omega = df_1 \wedge dx^1 + df_2 \wedge dx^2 + df_3 \wedge dx^3$$

$$= \left(\frac{\partial f_1}{\partial x^1}dx^1 + \frac{\partial f_1}{\partial x^2}dx^2 + \frac{\partial f_1}{\partial x^3}dx^3\right)\wedge dx^1 + \left(\frac{\partial f_2}{\partial x^1}dx^1 + \frac{\partial f_2}{\partial x^2}dx^2 + \frac{\partial f_2}{\partial x^3}dx^3\right)\wedge dx^2$$

$$+\left(\frac{\partial f_3}{\partial x^1}dx^1 + \frac{\partial f_3}{\partial x^2}dx^2 + \frac{\partial f_3}{\partial x^3}dx^3\right)\wedge dx^3$$

$$= \left(\frac{\partial f_2}{\partial x^1} - \frac{\partial f_1}{\partial x^2}\right)dx^1\wedge dx^2 + \left(\frac{\partial f_3}{\partial x^2} - \frac{\partial f_2}{\partial x^3}\right)dx^2\wedge dx^3 + \left(\frac{\partial f_1}{\partial x^3} - \frac{\partial f_3}{\partial x^1}\right)dx^3\wedge dx^1$$

... (10.51.7)

Equation (10.51.5) in vector notation is $(\nabla\times\vec{F})\cdot\hat{n}\,ds$

Hence, using (10.51.3), (10.51.6) and (10.51.7) we get the Stokes' theorem (10.51.5) in R^3.

Gauss's Divergence Theorem in R^3 : Let $V \subset R^3$ be a volume bounded by a closed smooth surface S. If $\vec{F}(x,y,z)$ be a continuous vector function which has continuous first partial derivatives in $V \subset R^3$, then

$$\iiint_V \nabla \cdot \vec{F}\, dV = \iint_S \vec{F} \cdot \hat{n}\, ds \quad \text{... (10.51.8)}$$

where $\hat{n}$ is the outward drawn unit normal vector to surface *S*.

Using the differential form the above Gauss's divergence theorem can be put in compact form as,

$$\iiint_S d\omega = \iint_{\partial S} \omega \quad \text{... (10.51.9)}$$

where S, represents the interior of a closed surface in R^3 and ∂S represents its boundary.

Proof: In order to obtain the Gauss's divergence theorem from the integral theorem(10.51.3), taking

$$\omega = f_1\, dx^2 \wedge dx^3 + f_2\, dx^3 \wedge dx^1 + f_3\, dx^1 \wedge dx^2 \quad \text{... (10.51.10)}$$

(which in vector notation corresponds to $\vec{F} \cdot \hat{n}\, ds$) be a differential 2-form in R^3, where f_i's are functions of x^1, x^2, x^3, then

$$d\omega = \left(\frac{\partial f_1}{\partial x^1} + \frac{\partial f_2}{\partial x^2} + \frac{\partial f_3}{\partial x^3}\right) dx^1 \wedge dx^2 \wedge dx^3 \quad \text{... (10.51.11)}$$

(which in vector notation equals $\nabla \cdot \vec{F}\, dV$). $\nabla \cdot \vec{F}\, dV$

Hence, using equations (10.51.3), (10.51.10) and (10.51.11), we get the Divergence theorem (10.51.9) in R^3.

The above results can be generalized to n-dimension, as the equation (10.51.3) is also valid in higher dimensions and the exterior product of a $p-$form and a $q-$ form is $(p+q)-$ form and the exterior derivative of a $p-$form is a $(p+1)-$ form.

11

Sprays

(11.1) Let X be a C^{∞} manifold modelled on a Banach space R over A. Let (η, Y, X) be a vector bundle. Let a be any real number. Then h_a is defined to be the map of Y into itself such that for each $x \in X$ and each $y \in \eta^{-1}[x]$, $h_a(y) = ay$, which is of course in $\eta^{-1}[x] \cdot h_a$ is a vector bundle morphism of η into itself. If $a \neq 0$, then h_a is a vector bundle isomorphism of η onto itself.

Suppose that $Y = X \times S$, where S is a Banach space over A, and let pr_1 be the projection of $X \times S$ onto X. Then $(pr_1, X \times S, X)$ is a trivial bundle. Let h_a be the map $X \times S \ni (x, s) \mapsto (x, as) \in X \times S$. So Th_a will be the map on the tangent bundle $T(X \times S) = (X \times S) \times R \times S$ into itself. Hence for $(x, s) \in \{x\} \times S$ and $(r, s') \in R \times S$ we have

$$\begin{aligned} Th_a((x, s), (r, s')) &= (h_a(x, s), Dh_a(x, s) \cdot (r, s')) \\ &= ((x, as), Did_{\{x\}}(x) \cdot r, D(s \mapsto as)(s) \cdot (s')) \\ &= ((x, as), (id_{\{x\}}(r), as')) = ((x, as), (r, as')). \end{aligned}$$

(11.2) Let X be a C^{∞} manifold modeled on a Banach space R over A. A C^{∞} vector field v on the tangent bundle TX of X is called a C^{∞} second order differential equation over X iff for all $y \in TX$,

$$(T(\tau_X))(v(y)) = y \qquad \text{...(11.2.1)}$$

Clearly, a vector field v on TX is a C^∞ second order differential equation over X iff $T(\tau_X) \circ v = id_{TX}$ or $\tau_{TX} \circ v = id_{TX}$. A C^∞ second order differential equation v over X is called a C^∞ spray over X iff for all real number a and all $y \in TX$,

$$v \circ h_a (y) = (T(h_a))(a\, v(y)) \qquad ...(11.2.2)$$

or equivalently

$$v \circ h_a (y) = a\, (T (h_a))(v(y)) \qquad ...(11.2.3)$$

(11.3) Let X be an open subset of a Banach space R over A. Then $TX = X \times R$ and $T(TX) = (X \times R) \times R \times R$. A C^∞ vector field v on $X \times R$ is a second order differential equation over X iff for all (x, r) in $X \times R$, $T(\tau_X) \circ v(x, r) = (x, r)$ or $T(\tau_X)((x, r), v'(x, r)) = (x, r)$, where v' be a map on $X \times R$ onto $R \times R$ called the vector part of v given by $v'(x, r) = (r, v_2(x, r))$, where v_2 is a map on $X \times R$ into R. Thus

$$T(\tau_X)((x, r), v'(x, r)) = (\tau_X(x, r), D\tau_X(x, r).\, v'(x, r)) = (x, r) \qquad ...(11.3.1)$$

here $\tau_X = pr_1$, hence, we have $(pr_1(x, r), D\, pr_1(x, r).v'(x, r))$ $(x, pr_1(v'(x, r))) = (x, r)$. Hence a C^∞ vector field v on $X \times R$ is a second order differential equation on X iff for all (x, r) in $X \times R$,

$$v'(x, r) = (r, v_2(x, r)) \qquad ...(11.3.2)$$

A C^∞ second order differential equation v over X is called a Spray over X iff for each real number $a \in A$ and each $x \in X$ and each $r \in R$,

$$v_2(x, ar) = a^2\, v_2(x, r) \qquad ...(11.3.3)$$

In fact, for each (x, r) in $X \times R$, by the condition of Spray $v \circ h_a(x, r) = Th_a(a\, v(x, r))$ we have L.H.S. $v(x, ar) = ((x, ar), (ar, v_2(x, ar)))$,

$$\begin{aligned} \text{R.H.S. } Th_a(a((x, r), (r, v_2(x, r)))) &= a\, (Th_a((x, r), (r, v_2(x, r)))) \\ &= a\, (h_a(x, r), Dh_a(x, r).(r, v_2(x, r))) \\ &= a\, ((x, ar), (r, av_2(x, r))) \\ &= ((x, ar), (ar, a^2 v_2(x, r))). \end{aligned}$$

Equating L.H.S = R.H.S., we have

$$v_2(x, r) = a^2 v_2(x, r) \quad ...(11.3.4)$$

Let f be an integral solution for the second order differential equation v at a point (x, r) of $X \times R$. Then f is a C^∞ morphism of an open interval J of R with center 0 into $X \times R$, and so there exists unique C^∞ maps g and h of J into X and J into R respectively such that for each $t \in J$, $f(t) = (g(t), h(t))$, hence for each $t \in J$, $Df(t) = (Dg(t), Dh(t))$. But since f is an integral solution for v at (x, r), $Tf \circ \bar{1}(t) = (d/d(t))\, f(t) = v \circ f(t)$. We have from L.H.S.

$$Tf \circ \bar{1}(t) = Tf(t, 1) = (f(t), Df(t).1)$$
$$= ((g(t), h(t)), (Dg(t), Dh(t))) \quad ...(11.3.5)$$

Also, from R.H.S.,

$$v \circ f(t) = v(f(t)) = v(g(t), h(t))$$
$$= ((g(t), h(t)), v'(g(t), h(t)))$$
$$= ((g(t), h(t)), (h(t), v_2(g(t), h(t)))) \quad ...(11.3.6)$$

[Since v is second order differential equation].

Thus we have on equating equation(11.3.5) and (11.3.6),

$$Dg(t) = h(t) \text{ and } Dh(t) = v_2(g(t), h(t)),$$

Consequently, we have

$$Df(t) = (h(t), v_2(g(t), h(t))) \text{ and } Dh(t) = D^2 g(t)$$

(11.4) *Theorem*:

Let v be a C^∞ Spray over a separated C^∞ manifold X. For each $y \in TX$,

Let $f_y : J_y \mapsto TX$ be the maximal integral solution for v at y. Then for each $y \in TX$ and each pair of real numbers s and t, $t \in Dm(f_{sy})$ iff $st \in Dm(f_y)$ and then, $f_{sy}(t) = sf_y(st)$, moreover for each $y \in TX$, $(T(\tau_X \circ f_y)) \circ \bar{1} = f_y$

Proof: Let $y \in TX$ and let s be any fixed real number. Let t be any real number such that $st \in Dm(f_y)$. Let S be the map $t \mapsto st$, and put $g = h_S \circ f_y \circ S$. Then for all such t,

$$Tg \circ \bar{1}(t) = T(h_S \circ f_y \circ S) \circ \bar{1}(t)$$
$$= T(h_S \circ f_y) \circ TS \circ \bar{1}(t)$$

$$= T(h_S \circ f_y) \circ TS(t, 1)$$
$$= T(h_S \circ f_y)(s(t), Ds(t).1)$$
$$= T(h_S \circ f_y)(s(t), s)$$
$$= T(h_S \circ f_y)(s, \bar{1}(st))$$
$$= s.(T(h_S) \circ Tf_y \circ \bar{1}(st)))$$
$$= s.(T(h_S) \circ v \circ f_y(st))) \text{ [} f_y \text{ is an integral solution]}$$
$$= s.(T(h_S) \circ v \circ f_y \circ S(t))$$
$$= v \circ h_S(f_y \circ S(t)) \text{ [} v \text{ is an Spray over } X \text{]}$$
$$= v \circ h_S \circ f_y \circ (st)$$
$$= v \circ h_S \circ f_y \circ S(t)$$
$$= v \circ g(t) \text{ and } g(0)$$
$$= h_S \circ f_y \circ S(0)$$
$$= h_S(f_y(0))$$
$$= s(f_y(0))$$
$$= sy$$
$$= f_{Sy}(0)$$

Hence g is an integral solution for $v \Rightarrow g = f_{Sy}$.

Also
$$f_{Sy}(t) = g(t)$$
$$= h_S \circ f_y \circ S(t)$$
$$= h_S \circ f_y(st)$$
$$= s.f_y(st)$$

Moreover
$$T(\tau_X \circ f_y) \circ \bar{1}(t) = T(\tau_X) \circ Tf_y \circ \bar{1}(t)$$
$$= T(\tau_X) \circ v \circ f_y(t)$$
$$= (id_{TX})(f_y(t))$$
$$= f_y(t)$$
$$\therefore T(\tau_X \circ f_y) \circ \bar{1} = f_y$$

Conversely, Let $t \in Dm(f_{Sy})$.

Then to prove $st \in Dm(f_y)$.

Let $t \in Dm(f_{sy})$, then $(1/s)st \in Dm(f_{sy})$

$$\Rightarrow \quad st \in Dm(f_{(1/s)sy})$$
$$\Rightarrow \quad st \in Dm(f_y)$$

(11.5) Let v be a C^{∞} Spray over a separated C^{∞} manifold X and let f be the flow for v. Let $(TX)_v$ be the set of all $y \in TX$ such that $1 \in J_y$. Evidently $(TX)_v$ is non void, where J_y being the domain of the maximum flow for v at y.

Let y be any element of $(TX)_v$ and let t be any real number in $[0,1]$. Then f_y is the maximal integral solution for v at y and $t \in Dm(f_y)$. By (11.4), $f_{ty}(1) = t f_y(1)$. Hence $t \in Dm(f_{ty})$. Consequently $[0,1]$ is contained in $Dm(f_{ty})$ and so ty is in $(TX)_v$. By (9.),$(TX)_v$ is an open subset of TX. The map of $(TX)_v$ into X which assigns to each $y \in (TX)_v$ the value $\tau_X \circ f(1, y)$ in X is called the exponential map of $(TX)_v$ into X induced by v and is denoted by exp_v or simply by exp, exp is C^{∞}.

Let $x \in X$. Put $(T_x X)_v = (TX)_v \cap T_x X$ and $exp_x = exp|(T_x X)_v$. Then exp_x is called the exponential map at x induced by v. The zero vector 0_x in $T_x X$ is also in $(T_x X)_v$ and

$$exp_x(0_x) = x,\ exp_x(0_x) = \tau_X \circ f(1, 0_x) = \tau_X\left(f_{0_x}(1)\right) = \tau_X(0_x) = x.$$

(11.6) *Theorem*:

Let v be a C^{∞} Spray over a separated C^{∞} manifold X. Then for each $x \in X$, exp_x is a local diffeomorphism at the zero vector 0_x in $(T_x X)_v$.

Proof: Let $x \in X$, for each vector $y \in T_x X$, let I_y be an open interval of R with center O such that for all $t \in I_y$, ty is in $(T_x X)_v$. Let g_y be the map of I_y into $(T_x X)_v$ such that $g_y(t) = ty$ for all $t \in I_y$ and let $h_y = exp_x \circ i_x \circ g_y$, where i_x is the inclusion map of $(T_x X)_v$ into $(TX)_v$. Then for each $y \in T_x X$ and each $t \in I_y$,

$$\begin{aligned} h_y(t) &= exp \circ i_x \circ g_y(t) \\ &= exp \circ i_x(ty) \\ &= exp\,(ty) \\ &= \tau_X \circ f(1, ty) \\ &= \tau_X\left(f_{ty}(1)\right) \\ &= \tau_X\left(tf_y(1)\right) \\ &= \tau_X\left(f_y(t)\right) \qquad [\tau_X(sz) = \tau_X(z) \text{ for all } z \in T_x X], \end{aligned}$$

hence $h_y = \tau_X \circ f_{\cdot y}$, where f is the flow for v. Now by (11.4) for each $y \in T_x X$,

$$Th_y \circ \bar{1}(0) = f_{\cdot y}(0) = y$$

on the other hand for each $y \in T_x X$,

$$\begin{aligned}(Th_y) \circ \bar{1}(0) &= T(exp \circ i_x \circ g_y) \circ \bar{1}(0) \\ &= T(exp \circ i_x) \circ Tg_y \circ \bar{1}(0) \\ &= T(exp \circ i_x)(g_y(0), Dg_y(0).1) \\ &= T(exp \circ i_x)(0_x, y)\end{aligned}$$

Since $(T_x X)_v$ is an open subset of Banach space $T_x X$, we have $((d(\varphi^k \circ f))(f_y(0_x)); j \in 2\underline{n})\ T_{(x,x)}(X \times X)$

$$T_{0x}((TX)_v)\ T_{0x}((T_x X)_v) = \{0_x\} \times T_x X,\ id_{((T_x X)_v)}$$

is the only chart for $T_x X$. Hence

$$\left(id_{((T_x X)_v)}\right)_{0x} : T_{0x}((TX)_v) \mapsto T_x X \text{ , Thus, } (0_x, y) = \left(id_{((T_x X)_v)}\right)_{0x}^{-1}(y)$$

$$\text{Consequently } (Th_y) \circ \bar{1}(0) = T(exp \circ i_x) \circ \left(id_{((T_x X)_v)}\right)_{0x}^{-1}(y)$$

$$\text{or} \qquad y = T(exp \circ i_x) \circ \left(id_{((T_x X)_v)}\right)_{0x}^{-1}(y)$$

Hence

$$T(exp \circ i_x) = \left(id_{((T_x X)_v)}\right)_{0x}$$

which is a linear homeomorphism, but $exp \circ i_x = exp_x$, hence $T(exp_x)$ is a linear homeomorphism and so by the inverse mapping theorem exp_x is a local diffeomorphism.

(11.7) *Theorem*:

Let v be a C^∞ Spray over a separated C^∞ manifold X modeled on A^n. Then the function $f: y \mapsto (\tau_X(y), exp(y))$ on $(TX)_v$ into $X \times X$ is a local diffeomorphism at 0_x for each $x \in X$.

Proof: Let θ be a chart for X at a point x of X, let θ be the chart for TX associated with θ and let φ be the chart for $X \times X$ at (x, x) such that $Dm(\varphi) = Dm(\theta) \times Dm(\theta)$ and for all (p, q) in $Dm(\varphi)\varphi(p, q) = (\theta(p),\theta(q))$.

Let $(e_i: i \in 2\underline{n})$ (respectively $(f_j: j \in 2\underline{n})$, $(g_k: k \in 2\underline{n})$) be the list of the basis vector fields relative to θ (respectively θ,φ). Then by (9.4.2) for each $j \in 2\underline{n}$,

$$(Tf) \circ f_j(0_x) = \sum_{k \in 2\underline{n}} \left(d\left(\varphi^k \circ f\right)\right)\left(f_j(0_x)\right) g_k f(0_x)$$

Consider $Tf : T_{0_x}\left((TX)_v\right) \mapsto T_{f(0_x)}(X \times X)$.

Now by (8.5) for each $k \in \underline{n}$

$$\begin{aligned} \varphi^k \circ f(y) &= pr_k(\varphi \circ f)\,(y) \\ &= pr_k(\varphi(\tau_X(y), exp(y)) \\ &= \varphi^k \circ \tau_X(y) \\ &= \theta^k \circ \tau_X(y), \qquad [\varphi = \theta \text{ for } k \in \underline{n}] \\ &= \theta^k(y)\varphi^k \circ f \\ &= \theta^k \end{aligned}$$

and for each $(n+k) \in 2\underline{n}$, $\varphi^{n+k} \circ f = \theta^k \circ exp$.

Hence by (5.5) for each $k \in \underline{n}$ and each $j \in \underline{2n}$

$$(d(\varphi^k \circ f\,))\,(\,f_j(0_x)) = (d(\theta^k\,))\,(\,f_j(0_x)) = \delta_{jk}$$

For each $i \in \underline{n}$, Let h_i be the function $A \ni t \to te_i(x) \in T_xX$

Then for each $i \in \underline{n}$ as in the proof of (11.6),

$$(T(i_x \circ h_i)) \circ \bar{1}(0) = (Ti_x) \circ \left(id_{((T_xX)_v)}\right)^{-1}_{0_x}(e_i(x))$$

and by (9.4.2)

$$(T(i_x \circ h_i)) \circ \bar{1}(0) = \sum_{j \in \underline{2n}} \left(d\left(\theta^k \circ i_x \circ h_i\right)\right)\left(\bar{1}(0)\right)\left(f_j(i_x \circ h_i)\right)(0),$$

Where i_x is the inclusión map of $(T_xX)_v$ into $(TX)_v$.

Now by (8.5) for each $i \in \underline{n}$ and each $j \in \underline{n}$ and each $t \in A$,

$$\underset{\sim}{\theta^j} \circ i_x \circ h_i(t) = \theta^j(x) \text{ and}$$

$$\underset{\sim}{\theta}^{n+j} \circ i_x \circ h_i(t) = t(d\theta^j)(\theta_i(x)) = t\delta_{ij}$$

Hence for each $j \in \underline{n}$,

$$(\mathrm{T}(i_x \circ h_j)) \circ \bar{1}(0) = (f_{n+j}(0_x))$$

and so by (11.6),

$$(\mathrm{T}(exp))(f_{n+j}(0_x)) = (\mathrm{T}(exp)) \circ (\mathrm{T}(i_x \circ h_j)) \circ \bar{1}(0)$$

$$= (\mathrm{T}(exp \circ i_x \circ h_j)) \circ \bar{1}(0)$$

$$= e_j(x).$$

Hence by (5.5) for each $j \in \underline{n}$,

$$(\mathrm{T}(exp))(f_{n+j}(0_x)) = \sum_{j \in \underline{n}} \left(d\left(\theta^j \circ exp\right)\right)\left(f_{n+j}(0_x)\right)\left(e_i\left(exp(0_x)\right)\right)$$

$$= \sum_{i \in \underline{n}} \left(d\left(\theta^i \circ exp\right)\right)\left(f_{n+j}(0_x)\right) e_i(x)$$

and so for each $i, j \in \underline{n}$,

$$(d(\theta^i \circ exp))\,(f_{n+j}(0_x)) = \delta_{ij}$$

Summing up the above arguments, we can state that the matrix

$$((d(\varphi^k \circ f))\,(f_j(0_x)); (k, j) \in \underline{2n} \times \underline{2n})$$

of the linear map $T_{0_x}(f)$ relative to the basis

$$(g_k(x,x);\ k \in \underline{2n}) \text{ of } \mathrm{T}_{(x,x)}(X \times X)$$

and the basis $((f_j(0_x));\ j \in \underline{2n})$ of $T_{0_x}((TX)_v)$

is of rank 2n and as such invertible. Hence f is a local diffeomorphism at 0_x.

12
Lie Groups

A Lie group appears to be a somewhat unnatural marriage between the algebric concept of a group and the differential-geometric notion of a manifold. This combination of algebra and calculus leads to powerful techniques for the study of symmetry which are not available for finite group see gorenotein [].

The distinguishing feature of a Lie group is that it also carries the structure of a smooth manifold, so the group elements can be smoothly varied.

(12.1) ***Definition of topological group***:

A group G with a topology is called a topological group iff the group operation

$$G \times G \ni (x, y) \mapsto xy^{-1} \in G$$

is continuous relatived to the above topology.

(12.2) ***Definition***:

Let G be a topological group. Then for each g in G the function L_g

$$G \ni x \mapsto g\,x \in G$$

and the function R_g

$$G \ni x \mapsto x\,g \in G$$

are homeomorhisms of G onto itself.

(12.3) Let G be a topological group with identity e and let X be a topological space. Then a continuous function a:(g, x) $\mapsto$ g.x of $G \times X$ into X is called an action of G on *X* iff for each element x of X (gh).x = g.(h.x) and e.x = x. We say G acts or operates on X or that X is a G-set iff there exists an action of G on X.

Let W be a topological subspace of X such that for each g in G and each x in W g.x is in W. Then the function $G \times W \ni (g,x) \mapsto g \cdot x \in W$ is called the restriction of the action a to W.

Let a:(g, x) $\mapsto$ g.x be an action of G on X. Two elements x and y of X are said to be equivalent under a iff there is an element g of G such that g.x = y. The relation of equivalence under a is an equivalence relation in X, called the equivalence relation of a. The equivalence class of an element x of X relative to the equivalence relation of a is called the orbit of x under a and is denoted by G.x; evidently G.x is the set of all y in X such that for each y there is a g in G such that g.x = y.

We denote the set of the orbits of all elements of X under a by X/G. The function on X into X/G which assigns to each x in X its orbit G.x is called the cannonical map of X onto X/G. X/G furnished with the strongest topology rendering the canonical map of X onto X/G continuous is called the orbit space of a. The canonical map of X onto X/G is open.

Let x be any element of X. Let G_x be the set of all elements g of G such that g.x = x. Then G_x is a subgroup of G, and is called the isotropy or stability subgroup of G at x. Evidently

$$G_x = (a_{.x})^{-1}[x], \text{where } a_{.x}: g \mapsto g \cdot x.$$

(12.4) By a C^k ($k \in \underline{N}$) group is meant a set G which is a group and which also has the C^k manifold structure such that the group operation

$$G \times G \ni (x,y) \mapsto xy \in G$$

and inverse operation

$$G \ni x \mapsto x^{-1} \in G$$

are C^k morphisms. Actually there is no need of writing second condition since it can be derived from first by using implicit function theorem.

A C^∞ group is called a <u>smooth group</u>. A finite -dimensional separated smooth group is called a <u>Lie group</u>.

(12.5) ***Matrix Lie Group:***

A *Matrix Lie group* G is a non-empty set of invertible matrices with real or complex entries, together with the operation of matrix multiplication. The set G must be closed under the operation of matrix multiplication, matix inversion and taking limit with in the set of all invertible matrices.

The last criterion of the definition means that if $\{g_n\}$ is a sequence of matrices, each of which lies in G, and if the sequence converges (entry by entry) to an invertible matrice g, then g is an element of G.

Every matrix Lie group is a group. But every group is not a matrix Lie group.

For example (R, +) is a group but it is not a matrix Lie group because the group operation is not matrix multiplication. However, (R, +) is isomorphic to matrix Lie group

$$G := \left\{\begin{pmatrix} 1 & t \\ 0 & 1 \end{pmatrix} : t \in R\right\}.$$

Define a function from (R, +) to G by

$$t \mapsto \begin{pmatrix} 1 & t \\ 0 & 1 \end{pmatrix}$$

which is differentiable, since each entry of the matrix is a differentiable function of t, it is also injective and serjective and respect the group operations

$$0 \mapsto \begin{pmatrix} 1 & 0 \\ 0 & 1 \end{pmatrix}$$

and

$$t + u \mapsto \begin{pmatrix} 1 & t+u \\ 0 & 1 \end{pmatrix} = \begin{pmatrix} 1 & t \\ 0 & 1 \end{pmatrix}\begin{pmatrix} 1 & u \\ 0 & 1 \end{pmatrix}.$$

(12.6) ***Examples*****:**

1. Let G = R, the set of real numbers, is a group with addition

operation and at the same it is also a smooth manifold so it is a Lie group as the group operation is smooth.

2. The general linear group GL(n, R), the set of all invertiable $n \times n$ matrices with real entries is a group under matrix multiplication. It can be identified with the open subset GL(n) = {A| det A ≠ 0}of the space $M_{n\times n}$ of all $n \times n$ matrices. But $M_{n\times n}$ is isomorphic to R^{n^2}, the coordinate being the matrix entries a_{ij} of A. Thus GL(n) is also an n^2- dimensional manifold. Hence it is a Lie group as the group operation is smooth.
3. The group GL(V) of invertible linear transformations of an n-dimensional vector V space over the field K can be regarded as a Lie group in view of the isomorphism GL(V) ≅ GL(n).
4. Let G = R^n,with the obvious manifold structure, and let the group operation be vector addition $(x, y) \mapsto x + y$. The inverse of a vector x is -x. Both these operations are clearly smooth, So R^n is an n-parameter abelian Lie group.
5. Let G = SO(2), the group of rotation in the plane, i.e.,

$$G = \left\{ \begin{pmatrix} \cos\theta & -\sin\theta \\ \sin\theta & \cos\theta \end{pmatrix} \middle| 0 \le \theta < 2\pi \right\}$$

where θ denotes the angle of rotation. Note that we can identify G with the unit circle

$$S^1 = \{(\cos\theta, \sin\theta)|\ 0 \le \theta < 2\pi\}$$

in R^2, which serves to define the manifold structure on SO(2). Thus SO(2) is a Lie group.
6. The orthogonal group

$$O(2) = \{A \in GL(2) \mid A^T A = I\}$$

has the manifold structure of two disconnected copies of S^1, is a Lie group.

In general the group of orthogonal $n \times n$ matrices

$$O(n) = \{A \in GL(n)|\ A^T A = I\}$$

is an open subset of R^{n^2} defined by the n^2 equations $A^TA - I = 0$, involving the matrix entries a_{ij} of A. It can be shown that precisely (1/2) n (n – 1) of these equations, corresponding to the matrix entries on or above the diagonal, are independent and satisfy the maximal rank condition everywhere on O(n). Thus O(n) is a regular submanifold of GL(n) of dimension (1/2)n (n-1). Matrix multiplication and inversion are smooth maps when restricted to O(n), hence O(n) is a Lie group.

The special orthogonal group

$$SO(n) = \{A \in O(n) | \det A = +1\}$$

being the connected component of thne identity of the orthogonal group is, also an (1/2)n (n-1)-parameter Lie group.

7. The group T(n) of upper triangular matrices with all 1\`s on the main diagonal is an (1/2)n (n-1)- parameter Lie group. As a manifold T(n) can be identified with the Euclidean space $\underline{R}^{n(n-1)/2}$. Since each matrix is uniquely determined by its entries above the diagonal for example, in the case of T(3), we identifiy the matrix

$$\begin{bmatrix} 1 & x & z \\ 0 & 1 & y \\ 0 & 0 & 1 \end{bmatrix} \in T(3)$$

with the vector (x, y, z) in $\underline{R}^3$. T(2) is isomorphic to $\underline{R}$ but T(n) is not isomorphic to the abelian Lie group $\underline{R}^{n(n-1)/2}$. In case of T(3), the group operation is given by

$$(x, y, z).(x', y', z') = (x + x', y + y', z + z')$$

8. Let $\underline{R}$ denote the manifold of real numbers with the usual coordinate chart $id_R: \underline{R} \to \underline{R} : t \to t$ and define the group operation $*: \underline{R} \times \underline{R} \to \underline{R}$ by $(x, y) \mapsto (x^3 + y^3)^{1/3}$ then ($\underline{R}$, *) is a topological group but not a Lie group, since * is not analytic at (0, 0), relative to above coordinate chart.
9. The manifold $\underline{R}$ of real numbers with the analytic structure on $\underline{R}$ given by coordinate chart $\theta: \underline{R} \to \underline{R} : t \mapsto t^3$ and

define the group operation $* : \underline{R} \times \underline{R} \to \underline{R}$ by $(x, y) \mapsto (x^3 + y^3)^{1/3}$ then $(\underline{R}, *)$ is a Lie Group.

10. $(\underline{R}, .)$ is neither a group nor a matrix Lie group.
11. The Lie group $(\underline{R}, +)$ is isomorphic to the matrix Lie group $(\underline{R}^+, .)$ via the correspondence $(\underline{R}, +) \ni t \mapsto e^t \in \underline{R}^+$.
12. For any natural number n the set of invertible $n \times n$ matrices with rational entries is a group denoted by GL(n, Q) but it is not a matrix Lie group. For the reasion given below Consider the group GL(2, Q), let

$$\begin{pmatrix} 3 & 0 \\ 0 & 1 \end{pmatrix}, \begin{pmatrix} 3.1 & 0 \\ 0 & 1 \end{pmatrix}, \begin{pmatrix} 3.14 & 0 \\ 0 & 1 \end{pmatrix}, \begin{pmatrix} 3.141 & 0 \\ 0 & 1 \end{pmatrix}, \ldots$$

where the nth matrix has a_{11} equal to the $(n - 1)^{th}$ decimal approximation of π. This sequence is contained in GL(2, Q), but its limit,

$$\begin{pmatrix} \pi & 0 \\ 0 & 1 \end{pmatrix}$$

is not . So GL(2, Q) is not a matrix Lie group. Similarly GL(n, Q) is not a matrix Lie group.

(12.7) ***Morphism of C^k groups*:**

Let G and G' be two C^k groups. Then a function f on G into G' is called a <u>morphism</u> of the C^k group G into the C^k group G' iff f is a C^k morphism of the C^k manifold G into the C^k manifold G', and f is a morphism of the group G into the group G'.

(12.8) ***Examples*:**

1. The exponential map $x \mapsto e^x$ is a homomorphism from the additive group $\underline{R}$ to the multiplicative Lie group $\underline{R}$ of the field $\underline{R}$.
2. The map $A \mapsto \det A$ is a homomorphism from the Lie group GL(n) to the multiplicative Lie group $\underline{R}$.
3. For any element g of a Lie group G the inner automorphism Ina (g): $x \mapsto g x g^{-1}$ is a Lie group automorphism.

4. The map $x \mapsto e^{ix}$ is a homomorphism from the Lie group $\underline{R}$ to the Lie group $S^1 = \{z \in C : |z| = 1\}$.

(12.9) Let G be a C^k ($k \in \underline{N}_*$) group with identity e. Then for each g in G the function L_g

$$G \ni x \mapsto g\,x \in G \qquad (12.9.1)$$

the function R_g

$$G \ni x \mapsto xg \in G \qquad (12.9.2)$$

and the function Ina(g)

$$G \ni x \mapsto g\,x\,g^{-1} \in G \qquad (12.9.3)$$

are C^k diffeomorphisms of G, and for each x in G

$$T_xG \xrightarrow{T_xL_g} T_{gx}G$$

$$T_xG \xrightarrow{T_xR_g} T_{xg}G$$

and $$T_xG \xrightarrow{T(\mathrm{Ina}(g))} T_{gxg^{-1}}G \qquad (12.9.4)$$

are linear homeomorhisms.

By abuse of notation for all x and g in G and all v in T_xG we write

$$g.v = T_x(L_g).v \text{ and } v.g = T_x(R_g).v \qquad (12.9.5)$$

Then for all x,g,h in G and all v in T_xG,

$$(g.v).h = g.(v.h)\ (= g.v.h)$$

(12.10) Let G be a C^k ($k \in \underline{N}$) group

The function m

$$G \times G \ni (x, y) \mapsto x\,y \in G \qquad (12.10.1)$$

is a C^k submersion and for each (x, y) in $G \times G$ each v in T_xG, and each w in T_yG

$$(T_{(x,y)}m)(v,w) = x.w + v.y \qquad (12.10.2)$$

The function i

$$G \ni x \mapsto x^{-1} \in G \qquad (12.10.3)$$

is a C^k diffeomorphism, and for each x in G and each v in T_xG

$$T_x i.v = -x^{-1}.v.x^{-1} \qquad (12.10.4)$$

Proof: (12.10.2): By (6.6) for each (x, y) in G × G each v in T_xG, and each w in T_yG

$$(T_{(x,y)}(m)).(v,w) = T_y(L_x).w + T_x(R_y).v = x.w + v.y$$

For proving (12.10.4) : Let f be the composition of m and the function p

$$G \ni x \mapsto (x, i(x)) \in G$$

Then for each x in G

$$f(x) = m(x, i(x)) = xx^{-1} = e,$$

Hence for each x in G and each v in T_xG

$$T_xf.v = 0$$

But for each x in G and each v in T_xG

$$\begin{aligned} T_xf.v &= T_x(mop.v) \\ &= (T_{p(x)}(m))(T_xp.v) \\ &= (T_{p(x)}(m))(v,T_xi.v) \\ &= (T_{(x,x-1)}(m))(v,T_xi.v) \\ &= x.(T_xi,v) + v.x^{-1} \end{aligned}$$

Hence for each x in G and each v in T_xG

$$x.(T_xi.v) + v.x^{-1} = 0$$

and so $\qquad T_xi.v = - x^{-1}.v.x^{-1}$

(12.11) Let f be a morphism of a C^k ($k \in \underline{N}_*$) group G with identity e into a C^k group G′ with identity e′. Let x be any element of G. Then for all y in G

$$f(xy) = f(x) f(y)$$

Hence

$$f \circ L_x = L_{f(x)} \circ f$$

and so

$$\begin{aligned} T_xf \circ T_eL_x &= T_e(f \circ L_x) \\ &= T_e(L_{f(x)} \circ f) \\ &= T_{f(e)}L_{f(x)} \circ T_ef \\ &= T_{e'}L_{f(x)} \circ T_ef. \end{aligned}$$

Consequently T_xf is injective (respectively surjective) iff T_ef is injective (respectively surjective).

(12.12) The torus$\underline{T} = \underline{R} / \underline{Z}$ is a commutative topological group. The map

$$r \mapsto (\cos r, \sin r)$$

is a continuous surjective map of $\underline{R}$ onto the circle

$$S^1 = \{(\cos r, \sin r) \in \underline{R} \times \underline{R} \,;\, r \in \underline{R}\}$$

Hence the map

$$\underline{R} / \underline{Z} \ni r + \underline{Z} \mapsto (\cos r, \sin r) \in S^1$$

is a homeomorphism of the topological space $\underline{T}$ onto the topological space S^1, called <u>canonical</u>. $\underline{T}$ becomes a Lie group, when the structure of the smooth manifold S^1 is transported to $\underline{T}$ by the canonical homeomorphism. Then the canonical map

$$\underline{R} \ni r \mapsto r + \underline{Z} \in \underline{R} / \underline{Z}$$

is an epimorphism of the Lie group $\underline{R}$ onto the Lie group $\underline{T}$.

(12.13) Let G be a Banach algebra over A. Then (by Linear analysis) the set X_* of all invertible elements of X is an open subset of X, and as such an open submanifold of the manifold X. By (Local Analysis) the group operation

$$X \times X \ni (x, y) \mapsto xy \in X$$

is smooth (By Local Analysis) the function

$$X \ni x \mapsto x^{-1} \in X$$

is smooth. Thus X is a Lie group modelled on the Banach space X over A.

(12.14) Let X be a Banach space over A. Then the set L(X, X) of all continuous linear maps of X into X is a Banach algebra over A. Hence (By Linear analysis) the set GL(X) of all linear homeomorphisms of X onto X is a smooth group. GL(X) is a smooth group, modelled on the Banach algebra L(X, X). We identify the tangent space at the identity id_X in GL(X) with L(X, X). Next suppose that $X = \underline{R}^n$, for some positive integer n. Then GL(X) can be identified with the general linear group in n variables over $\underline{R}$, GL(n, $\underline{R}$), consisting of all invertible $n \times n$ matrices in $\underline{R}$. With this identification GL(n, $\underline{R}$) is a Lie group, modelled on the Banach space $\underline{R}^{n^2}$ of all square matrices of degree n over $\underline{R}$. $\mathrm{Dim}_A\,(GL(n, \underline{R})) = n^2$.

Then the special linear group SL(n, $\underline{R}$) of degree n is a closed submanifold of the manifold GL(n, $\underline{R}$). Hence SL(n, $\underline{R}$) is a closed Lie subgroup of the Lie group GL(n, $\underline{R}$).

(12.15) ***Definition (12.15.1)***: Let G be a real (respectively complex) Lie group with Lie algebra $\mathfrak{g}$. Suppose V is a finite dimensional vector space over R (respectively on C), then by a representation of G in V we mean an analytic homomorphism of G into GL(V), where GL(V) denote the group of linear transformation from V onto V. It is obvious that if π is a representation of G in V, then $(x, v) \mapsto \pi(x)\, v$ is an analytic map of $G \times V$ intoV.

(12.15.2) Definition: Let G be a matrix Lie group. Then a finite-dimensional complex representation of G is a Lie group homomorphism

$$\pi: G \to GL(n; C)$$

$(n \geq 1)$ or, more generally, a Lie group homomorphism

$$\pi: G \to GL(V)$$

where V is a finite dimensional complex vector space (with dim $(V) \geq 1$). A finite dimensioanl real representation of G is Lie group homomorphism π of G into GL(n; R) or into GL(V), where V is a finite dimensional real vector space.

One should think of a representation as a Linear action of a group or Lie algebra on a vector space (say, to every $g \in G$, there is associated an operator $\pi(g)$, which acts on the vector space V).

(12.15.3) Definition : Let π be a finite - dimensioanl real or complex repesentation of a matrix Lie group G, acting on a space V. A subspace W of V is called invariant if $\pi(A)w \in W$ for all $w \in W$ and all $A \in G$. An invariant subspace W is called nontrivial if $W \neq \{0\}$and $W \neq V$. A representation with no nontrivial invariant subspace is called irreducible.

13
Lie Algebras of Lie Groups

Infinitesimal symmetries are Lie algebras mathematically, one can think of a Lie algebra g as the set of possible derivatives of parameterized paths at the unit element 1 of a Lie group G. Consider any parameterized path g(t) in the group that passes through the identity element at time 0, i.e., unit g(0) = 1. Then G′(0) is an element of the Lie algebra g, and each element of the Lie algebra g can be found as such a derivative. Physicists often work with derivatives. Suppose x and y depend on time t after a short time dt we write for values of x and y as x + dx and y + dy respectively. Our question is what is the value of xy after a short time dt? Since

$$(x + dx)(y + dy) = xy + (x\,dy + y\,dx) + \ldots$$

by ignoring the higher order terms in the infinitesimals, we have

$$\frac{d}{dx}(xy) = x\frac{dy}{dt} + y\frac{dx}{dt}$$

so one can think of Lie algebra elements as first-order terms in the Taylor series expansion of group element close to identity. The elements of g one generally called infinitesimal symmetries or infinitesimal generators of the group G.

Lie algebras of translation symmetry group and rotational symmetry plays important role in the analysis of two body problems.

(13.1) *Definition*:

Let X be a separated smooth group, modelled on a real Banach Space, with identity e. Let V be a vector space over A, then V is called a Lie algebra with bracket operatiomn [,]: $V \times V \mapsto V$ defined as $(v,v') \rightarrow [v.v']$ iff the vector product is bilinear, for each v in V, $[v.v] = 0$ and for each triple of vectors v, v', v'' the Jacobi identity

$$[v.[v'.v'']] + [v'.[v''.v]] + [v''.[v.v']] = 0.$$

is satisfied. The bracket [,] is called the Lie-bracket.

Let V and V' be Lie algebra over A with multiplications denoted by juxta-position. Then a function f on V into V' is called a morphism of the Lie algebra V into the Lie algebra V' iff f is linear and for any elements x and y of V,

$$f(x\,y) = f(x)\,f(y).$$

(13.2) *Definition*:

A vector field v on Lie-group X is said to be left (respectively right) invariant iff

$$(TL_x) \circ v \circ L_x^{-1} = v \text{ (respectively } (TR_x) \circ v \circ R_x^{-1} = v) \qquad (13.2.1)$$

for all x in X. The set of all left invariant vector fields on X is denoted by $La(X)$. $La(X)$ is a vector space over R, since for all x in X, and all v and w in $La(X)$

$$(TL_x) \circ [v, w] \circ L_x^{-1} = [(TL_x) \circ v \circ L_x^{-1},$$
$$(TL_x) \circ w \circ L_x^{-1}] = [v, w] \qquad (13.2.2)$$
$$\Rightarrow [v, w] \in La(X)$$

so $La(X)$ is a Lie-algebra over R with respect to Lie-bracket operation defined by

$$(v, w) \mapsto [v, w]$$

$La(X)$ is called the Lie sub-algebra of $\Gamma^k(X)$, the set of all vector fields on X

(13.3) *Theorem*:

For each v in $T_e X$ the function F_v

$$X \ni x \mapsto (TL_x)(v) \in TX$$

is a left invariant vector field on X, and the function F

$$T_e X \ni v \mapsto F_v \in La(X)$$

is a linear isomorphism, called canonical and so also its inverse.

Proof: Let x be any element of X and let v be any vector in $T_e x$. Then for each y in X

$$\begin{aligned}(TL_x)\circ F_v \circ L_x^{-1}(y) &= (TL_x)\circ F_v \circ L_{x^{-1}}(y)\\ &= (TL_x)\circ F_v\,(x^{-1} y)\\ &= (TL_x)\,((T L_{x^{-1}y})\,(v))\\ &= (T(L_x \circ L_{x^{-1}y}))(v)\\ &= (TL_y)(v)\\ &= F_v(y)\end{aligned}$$

Hence F_v is left invariant vector field and as such is in $La(X)$. Let v and w be any elements of $T_e X$ such that $F_v = F_w$.

Evidently $F_v(e) = v$ and $F_w(e) = w$, and so $v = w$. Consequently F is injective. Let z be any element of $La(X)$. Then for each x in X

$$(TL_x)\circ z \circ L_x^{-1} = z$$

and so

$$F_{z(e)}(x) = (TL_x)(z(e)) = z(L_x(e)) = z(x).$$

Hence for each element z of $La(X)$ there is a unique element $z(e)$ of $T_e X$ such that $F_{z(e)} = z$. The linearity of F is obvious. Thus $v \mapsto F_v$ is a linear isomorphism of $T_e X$ onto $La(X)$ and its inverse is the linear isomorhism $z \mapsto z(e)$ of $La(X)$ onto $T_e X$.

(13.4) Whenever $La(X)$ is considered as a Banach space, it is to be given the Banach space structure transformed by the canonical map of the Banach space $T_e X$. On the other hand $T_e X$ is taken to be a Lie algebra such that for two elements u and v of $T_e X$ $[u, v] = [u', v'](e)$ where u' and v' are elements of $La(X)$ such that $u'(e) = u$ and $v'(e) = v$. Thus $T_e X$ and $La(X)$ are both Banach space and Lie algebra over A. They are often identified by means of their canonical maps.

For each x in X, L_x is a diffeomorphism of X onto itself, therefore $T_e L_x$ is an isomorphism of $T_e X$ onto $T_x X$. Hence the map defined by $(x, v) \mapsto (T_e(L_x))(v)$ is an isomorphism of the trivial vector bundle

$X \times T_e X$ onto the tangent bundle TX, and so the tangent bundle TX is isomorphic with the trivial vector bundle $X \times La(X)$.

(13.5) *Theorem*:

Every left invariant vector field on X is globally integrable.

Proof: Let v be a left invariant vector field on X. Then there exists a positive real number a, an open neighbourhood U of the identity element e of X and a morphism g of $[-a, a] \times U$ into X such that g is a flow for v at e. Let x be any element of X, and let h be the function

$$[-a, a] \times L_x [U] \ni (t, z) \to L_x (g (t, L_x^{-1}(z))) \in X$$

Then by (9.12) h is a flow for $(TL_x) \circ v \circ L_x^{-1}$ at $L_x(e) = x$. But by hypothesis $(TL_x) \circ v \circ L_x^{-1} = v$. Hence v is globally integrable.

(13.6) *Theorem*:

Let f be the flow for a smooth left invariant vector field v on X. For each t in $\underline{R}$ let f_t be the diffeomorphism

$$X \ni x \mapsto ft. (x) = f(t, x) \in X$$

Then for all t in $\underline{R}$ and all x in X

$$Lx \circ ft. = ft. \circ Lx \tag{13.6.1}$$

Let h_v be the function

$$R \ni t \mapsto ft. (e) = f(t, e) \in X \tag{13.6.2}$$

Then h_v is a smooth homeomorphism of the additive Lie group $\underline{R}$ into the Lie group X such that for all t in $\underline{R}$ and for all x in X

$$R_{h_v(t)} = ft. \text{ and } xh_v(t) = f(t, x)$$

Furthermore h_v is the integral solution for v at e.
Finally for each x in X

$$v(x) = \left(\frac{d}{dt}(xh_v(t))\right)_{t=0} \tag{13.6.3}$$

Proof: By (9.18) for all t in $\underline{R}$ and all x in X

$$Lx \circ ft. = ft. \circ Lx$$

therefore

$$R_{h_v(t)}(x) = xh_v(t) = xft.(e) = ft.(xe) = ft.(x)$$

and so $\qquad R_{h_v(t)} = ft.$

For all s and t in R

$$\begin{aligned} h_v(s+t) &= f(s+t, e) \\ &= f(s, f(t, e)) \\ &= fs.(ft.(e)) \\ &= fs.(ft.(e).e) \\ &= fs.(\text{L}ft.(e)(e)) \\ &= \text{L}ft.(e)(fs.(e)) \\ &= ft.(e).fs.(e) \\ &= h_v(t).h_v(s) \end{aligned}$$

Thus h_v is a homomorphism of R into X.

Evidently $\quad h_v(o) = fo.(e) = f(o, e) = e$

Moreover for all t in R $\dfrac{dh_v(t)}{dt} = \dfrac{dft.(e)}{dt} = v(ft.(e)) = v(h_v(t)).$

Hence h_v is the integral solution for v at e. Let x be any element of X. Then for each t in R

$$f_t(x) = R_{h_v(t)}(x) = x\,h_v(t)$$

and so

$$v(ft.(x)) = \frac{dft.(x)}{dt} = \frac{d}{dt}(xh_v(t)).$$

Hence

$$v(x) = \left(\frac{d}{dt}(\text{xh}_{\text{v}}(\text{t}))\right)_{\text{t}=0}.$$

(13.7) Let f be a smooth homomorphism of the additive Lie group $\underline{R}$ into X. Then for all s and t in $\underline{R}$

$$f(s+t) = f(s)f(t) = L_{f(s)}(f(t)) = f(t)f(s) = R_{f(s)}(f(t))$$

and so keeping s fixed and varying t

$$\begin{aligned} \dot{f}(s+t) &= (T_{f(t)}(L_{f(s)}))(\dot{f}(t)) \\ &= (T_{f(t)}(R_{f(s)}))(\dot{f}(t)) \end{aligned}$$

$$= (T_{f(t)}(R_{f(s)}))(\dot{f}(t))$$

Hence, Letting t be 0,

$$\dot{f}(s) = (T_e(L_{f(s)}))(\dot{f}(o))$$

$$= (T_e(R_{f(s)}))(\dot{f}(o)) \qquad (13.7.1)$$

(13.8) *Theorem*:

Let $L(R, X)$ be the set of all smooth homomorphisms of the additive Lie group R into X. Let h be the function on $La(X)$ into $L(R, X)$ such that h assigns to each member v in $La(X)$ the member h_v of $L(R, X)$, the integeral solution for v at e. Then h is a bijection, called canonical.

Proof: Let f be any member of $L(R, X)$. Let v be a member of $La(X)$ such that $v(e) = \dot{f}(o)$. Then by (13.7) for all t in $\underline{R}$

$$\dot{f}(t) = (T_e(L_{f(t)}))(\dot{f}(o))$$

$$= (T_e(L_{f(t)}))\ v(e)$$

$$= v(L_{f(t)}\ e) \text{ (since } v \text{ is left invariant)}$$

$$= v(f(t)).$$

since f is a homomorphism, $f(0) = e$. Hence by (13.6) $h_v = f$. Thus h is surjective.

Let v and w be any two member of $La(X)$ such that $h_v = h_w$. Then $v(e) = \dot{h}_v(o) = \dot{h}_w(o) = w(e)$,

Hence $v = w$. Thus h is injective. We have thus shown that h is a bijection of $La(X)$ onto $L(R, X)$.

13.9) *Theorem*:

Let v and w be any member of $La(X)$. Let v' and w' be the member of $L(R, X)$ which are assigned to v and w respectively by the canonical bijection of $La(X)$ onto $L(R, X)$, let f and g be the flows for v and w respectively. Then for all s and t in $\underline{R}$

$$g_{t.} \circ f_{s.} = f_{s.} \circ g_{t.} \Leftrightarrow (w'(t))(v'(s)) = (v'(s))\,(w'(t)).$$

Proof: Let s and t be any member of $\underline{R}$ and let x be any member of X. Then by (13.6.1)

$$(f_{s.} \circ g_{t.})(x) = (f_{s.} \circ g_{t.} \circ L_x)(e) = (L_x \circ f_{s.} \circ g_{t.})(e)$$

But by (13.6.2)

$$g_{t.}(e) = w'(t) = (w'(t)).e = L_{w'(t)}(e)$$

Hence

$$\begin{aligned}(L_x \circ f_{s.} \circ g_{t.})(e) &= (L_x \circ f_{s.} \circ L_{w'(t)})(e) \\ &= (L_x \circ L_{w'(t)} \circ f_{s.})(e) \\ &= (L_x(L_{w'(t)}(f_{s.}(e)))) \\ &= L_x(L_{w'(t)}(v'(s))) \\ &= x(w'(t))\,(v'(s)).\end{aligned}$$

Consequently

$$(f_{s.} \circ g_{t.})(x) = x(w'(t))(v'(s)).$$

By similar argument

$$(g_{t.} \circ f_{s.})(x) = x(v'(s))(w'(t))$$

Hence

$$g_{t.} \circ f_{s.} = f_{s.} \circ g_{t.} \Leftrightarrow (w'(t))(v'(s)) = (v'(s))\,(w'(t)).$$

(13.10) Let x be any element of X, and put

$$\underline{Ad}(x) = T_e(Ina(x))$$

since $Ina(x)$ is a diffeomorphism of X, the function

$$\underline{Ad}(x)\colon v \mapsto (T_e(Ina(x)))(v) \tag{13.10.1}$$

is a linear homeomorphism of Banach space T_eX onto itself.

For any two elements x and y of X

$$\underline{Ad}(xy) = (\underline{Ad}(x)) \circ (\underline{Ad}(y)) \tag{13.10.2}$$

$$\underline{Ad}(e) = id_{T_eX} \tag{13.10.3}$$

and so

$$(\underline{Ad}(x))^{-1} = \underline{Ad}(x^{-1}) \tag{13.10.4}$$

The function

$$X \times T_eX \ni (x,v) \mapsto (\underline{Ad}(x))(v) \in T_eX \tag{13.10.5}$$

is C^{∞} and as such a C^{∞} action of the Lie group X on the Banach space $T_e X$.

Again let x be any element of X and let v be any element of $La(X)$ put

$$(Ad(x)(v)) = T(Ina(x)) \circ v \circ (Ina(x))^{-1} \tag{13.10.6}$$

Since

$$L_x \circ R_{x^{-1}} = Ina(x),$$

$$(Ad(x))(v) = (T(L_x \circ R_{x^{-1}})) \circ v \circ (L_x \circ R_{x^{-1}})^{-1} \tag{13.10.7}$$

Since for any two elements y and z of X

$$L_y \circ R_z = R_z \circ L_y,$$

$$\begin{aligned}(Ad(x))(v) &= TL_x \circ TR_{x^{-1}} \circ v \circ R_x \circ L_{x^{-1}} \\ &= TR_{x^{-1}} \circ TL_x \circ v \circ L_{x^{-1}} \circ R_x \\ &= TR_{x^{-1}} \circ v \circ R_x .\end{aligned} \tag{13.10.8}$$

Hence $(Ad(x))(v)$ is left invariant and as such belongs to $La(X)$.

Thus $Ad(x)$ is a function on $La(X)$ into itself. Let b be the canonical bijection of $La(X)$ onto $T_e X$. Then for each v' in $La(X)$

$$\begin{aligned}(\underline{Ad}(x)(b(v`))) &= (\underline{Ad}(x))(v'(e)) \\ &= (T_e(Ina(x)))(v'(e)) \\ &= (T_e(Ina(x)))(v((Ina(x))^{-1})(e)) \\ &= ((Ad(x))(v'))(e) \\ &= b((Ad(x))(v')).\end{aligned} \tag{13.10.9}$$

Hence by (13.4) and (13.10.5) the function

$$X \times La(X) \ni (x,v) \to (Ad(x))(v) \in La(x) \tag{13.10.10}$$

is a C^{∞} action of the C^{∞} group X on the Banach space $La(X)$. Consequently for each x in X $Ad(x)$ is a linear homeomorphism of $La(X)$ onto itself, and so the function $Ad : x \to Ad(x)$ is a morphism of the group X into the group of linear homeomorphism of $La(X)$ relative to composition and is called adjoint representation of X.

By (9.16) for each element x in X and each pair of elements v and w in $La(X)$

$$(Ad(x))\,[v, w] = [(Ad(x))(v), (Ad(x))(w)]$$

so that $Ad(x)$ is an automorphism of the Lie algebra $La(X)$.

(13.11) The contragredient of the adjoint representation $\underline{Ad}$ (respectively Ad) of X is called the coadjoint representation of X and is denoted by $\underline{Cad}$ (respectively Cad). Cad is a representation of X into the ***strong dual*** $(La(X))^*$ of $La(X)$. For each x in X $Cad(x) = (Ad(x^{-1}))^*$, the transpose of $Ad(x^{-1})$.The action of X on $La(X)$ (respectively $(La(X))^*$) induced by Ad (respectively Cad) is called the ***adjoint*** (respectively ***coadjoint***) action of X on $La(X)$ (respectively $(La(X))^*$) and is denoted by Ad (respectively Cad) by abuse of notation.

For each x in X, each v in $La(X)$ and each z in $(La(X))$

$$< v, (Cad(x))(z) > = < v, (Ad(x^{-1}))^*(z) > = < (Ad(x^{-1}))v, z > .$$

(13.12) Let f be a morphism of X into a separated C^∞ group X' with identity e' modelled on a real Banach space. Let v be any element of the algebra $La(X)$. Then $v(e)$ is an element of $T_e X$ and so $(T_e f)(v(e))$ is an element of $T_{e'} X'$.

For each element x of X

$$L_{f(x)} \circ f = f \circ L_x \qquad (13.12.1)$$

Let v' be an element of $La(X')$ such that

$$v'(e') = (T_e f)(v(e)).$$

Then for each x in X

$$\begin{aligned} v'(f(x)) &= (TL_{f(x)})(v'(e')) \\ &= (TL_{f(x)})((T_e f)(v(e))) \\ &= (T(L_{f(x)} \circ f))(v(e)) \\ &= (T(f \circ L_x))(v(e)) \\ &= (Tf)((TL_x)(v(e))) \\ &= (Tf)(v(x)) \end{aligned}$$

Hence

$$v' \circ f = (Tf) \circ v \qquad (13.12.2)$$

Let w be another element of $La(X)$ and let w' be an element of $La(X')$ such that $w'(e') = (Tf)(w(e))$.

Then by (9.16)

$$(Tf\ ) \circ [v, w] = [v', w'] \qquad (13.12.3)$$

Consequently the function on $La(X)$ into $La(X')$, which assigns to each v in $La(X)$ the elements v' in $La(X')$ such that

$$v' \circ f = (Tf\ ) \circ v$$

is a morphism of the Lie algebra $La(X)$ into the Lie algebra $La(X')$, is called the transfer of f to Lie algebra and is denoted by $La(f)$.

If b (respectively b') is the canonical bijection of $La(X)$ onto T_e X (respectively $La(X')$ onto $T_e X'$) then

$$(T_e f\ ) \circ b = b' \circ (La(f)) \qquad (13.12.4)$$

(13.13) *Linear Lie algebra of a closed linear group:*

A closed linear group G is a group of real or complex matrices that is topologically closed in a complex general linear group. Rotation groups, Unitary groups and Special linear groups provide familiar examples. The linear Lie algebra g of G is the set of derivatives at 0 of all smooth curves $c(t)$ of matrices that lie in G for all t and are equal to the identity at $t = 0$. The set of matrices g is indeed a Lie algebra over $\underline{R}$. The exponential of a square matrix is defined by the power series of the exponential function . The exponential map enables us to compute explicitly the linear Lie algebra of each of the familiar examples. It turns out that the exponential map carries g into G.

We denote by $GL(n, \underline{R})$ the real ***general linear group*** consisting of all nonsingular *n-by-n* real matrix with matrix multiplication as group operation. Similarly $GL(n, C)$ denotes the group of nonsingular *n-by-n* complex matrices. The groups $GL(n, \underline{R})$ and $GL(n, C)$ have topologies if we identify them with subset of $\underline{R}^{n^2}$ and $\underline{R}^{2n^2}$, respectively; in fact, each of these groups is an open subset in its Euclidean space, being the subset where the polymomial function *det* is not zero. Multipliction and inversion are continuous because they are given by polynomials in the entries and by division by the polynomial det, and thus $GL(n, R)$ and $GL(n, C)$ are topological groups.

We shall refer to any closed subgroup of some $GL(n, C)$ as a

closed linear group. The groups $GL(n, \underline{R})$ and $GL(n, C)$ are themselves closed linear groups. Any closed linear group inherits a topology from $GL(n, R)$ or $GL(n, C)$ and becomes a topological group. A few more examples of closed linear groups are the following:

$$
\begin{aligned}
SO(n) &= \{x \in GL(n, R) \mid xx^{T} = 1 \text{ and } det\, x = 1\} \\
U(n) &= \{x \in GL(n, C) \mid xx^{*} = 1\} \\
SU(n) &= \{x \in U(n) \mid det\, x = 1\} \\
SL(n, R) &= \{x \in GL(n, R) \mid det\, x = 1\} \\
SL(n, C) &= \{x \in GL(n, C) \mid det\, x = 1\} \qquad (13.13.1)
\end{aligned}
$$

The first three are the ***rotation group***, the ***unitary group***, and the ***special unitary group***; the last two are the ***special linear groups*** over R and C. The reason that these subgroups are closed is that they are defined by polynomial equations. The polynomials in question are the *entry-by-entry* relation in the real and imginary parts of the matrix entries that amount to the defining equation for the groups.

The study of closed linear groups as Lie groups begins from the fact that a kind of differentiation is avalable in any closed linear group. For any such group G, we can speak of ***smooth curves*** in G namely C^{∞} curves $c(t)$ into the underlying matrix space whose image is in G for each t. We are especially interested in smooth curves in G with the property that $c(0)$ is the identity 1. An example is the curve

$$c(t) = \begin{pmatrix} \cos t & \sin t & 0 \\ -\sin t & \cos t & 0 \\ 0 & 0 & 1 \end{pmatrix}$$

in the group $G = SO(3)$. For any such curve we can form $c'(0)$, which will be some 3-*by*-3 matrix . With c as in this example,

$$\dot{c}(t) = \begin{pmatrix} -\sin t & \cos t & 0 \\ -\cos t & -\sin t & 0 \\ 0 & 0 & 0 \end{pmatrix} \text{ and } \dot{c}(0) = \begin{pmatrix} 0 & 1 & 0 \\ -1 & 0 & 0 \\ 0 & 0 & 0 \end{pmatrix}$$

Here $\dot{c}(0)$ tells us the direction in G in which the curve extends from 1. We may think of it as telling how a parameter describing c is to start out from the identity. Let

$$\mathfrak{g} = \{ \dot{c}(0) \mid c: \underline{R} \to G \tag{13.13.2}$$

is a curve with $c(0) = 1$ that is smooth as a function into matrices}

Then $\mathfrak{g}$ is a subset of matrices of the same size as the underlying general linear group; the members of $\mathfrak{g}$ are not necessarily invertible.

The set $\mathfrak{g}$ in (13.13.2) has a certain amount of structure. First, $\mathfrak{g}$ is a real vector space. To see closure under addition, let $\dot{c}(0)$ and $\dot{b}(0)$ be in $\mathfrak{g}$. Forming $c(t)\, b(t)$, we have

$$\frac{d}{dt}[c(t)\, b(t)] = c(t)\, \dot{b}(t) + \dot{c}(t)\, b(t)$$

$$\frac{d}{dt}[c(t)\, b(t)]\Big|_{t=0} = c(0)\, \dot{b}(0) + \dot{c}(0)\, b(0)$$

$$= \dot{b}(0) + \dot{c}(0).$$

Thus $\mathfrak{g}$ is closed under addition. To see closure under scalar multiplication, let $\dot{c}(0)$ be in $\mathfrak{g}$ and let k be in $\underline{R}$. Forming $c(kt)$, we have

$$\frac{d}{dt}[c(kt)] = \dot{c}(kt)k \text{ and}$$

$$\frac{d}{dt}[c(kt)]\Big|_{t=0} = k\,\dot{c}(0)$$

Thus $\mathfrak{g}$ is closed under scalar multiplication.

Second, G is closed under group conjugation $x \to gxg^{-1}$. Thus if $c(t)$ is a smooth curve in G passing through 1 at $t = 0$, then so is $gc(t)g^{-1}$. Differentiating at $t = 0$, we see that $g\dot{c}(0)g^{-1}$ is in $\mathfrak{g}$. Thus $\mathfrak{g}$ is closed under the linear maps $Ad(g)$, for $g \in G$, that are given by

$$Ad(g)\, X = g\, X\, g^{-1} \tag{13.13.3}$$

Third, let X be in $\mathfrak{g}$ and let $c(t)$ be a smooth curve in G with $c(0) = 1$. Then $t \to Ad(c(t))\, X$ is a smooth function into $\mathfrak{g}$, as we see by substituting $g = c(t)$ into (13.13.3) and writing matters out. By definition we have

$$\frac{d}{dt} Ad(c(t))\, X\Big|_{t=0} = \lim_{t \to 0} \frac{1}{t}[Ad(c(t))\, X - X] \tag{13.13.4}$$

and we know that the left side exists. The right side involves vector space operation and a passage to limit, all on member of g. Since g, as a real vector subspace of matrix space, is closed topologically, the limit is in g. Let us calculate this limit

To brove a preliminary formula, namely

$$\frac{d}{dt}\, c(t)^{-1} = -\, c(t)^{-1}\, \dot{c}(t)\, c(t)^{-1} \tag{13.13.5}$$

we differentiate $c(t)\, c(t)^{-1} = 1$ by the product rule, being careful about the order in which we write down products. We obtain

$$\dot{c}(t)\, c(t)^{-1} + c(t)\, [\tfrac{d}{dt}\, c(t)^{-1}] = 0$$

so (13.13.5) follows.

Therefore

$$\begin{aligned} \frac{d}{dt}\, Ad(c(t))\, X &= \frac{d}{dt}\, [c(t)\, X\, c(t)^{-1}] \\ &= \dot{c}(t)\, X\, c(t)^{-1} + c(t)\, X\, [\tfrac{d}{dt}\, c(t)^{-1}] \\ &= (t)\, X\, c(t)^{-1} - c(t)\, X\, c(t)^{-1}\, \dot{c}(t)\, c(t)^{-1} \end{aligned}$$

Putting $t = 0$ and taking (13.13.4) into account, we see that

$$\dot{c}(0)\, X - X\, \dot{c}(0)$$

is in g. We conclude that g is closed under the Lie bracket operation

$$[X, Y] = XY - YX \tag{13.13.6}$$

The Lie bracket operation in (13.13.6) is linear in each variable and has the properties that

- (a) $[X, Y] = 0$ for all $X \in g$ (and hence $[X, Y] = -[Y, X]$)and
- (b) the Jacobi identity holds:

$$[[X, Y], Z] + [[Y, Z], X] + [[Z, X], Y] = 0$$

Property • (a) is clear, and property • (b) follows from the calculation

$$\begin{aligned} &[[X, Y]. Z] + [[Y, Z], X] + [[Z, X], Y] \\ &= [XY - YX, Z] + [YZ - ZY, X] + [ZX - XZ, Y] \\ &= (XYZ - YXZ) - (ZXY - ZYX) + (YZX - ZYX) \\ &\qquad - (XYZ - XZY) + (ZXY - XZY) - (YZX - YXZ) \\ &= 0 \end{aligned}$$

each variable and satisfies • (a) and • (b) is called a real ***Lie Algebra***. Accordingly we call $\mathfrak{g}$ in (13.13.2) the ***linear Lie algebra*** of the closed linear group G.

Let $gl\,(n, \underline{R})$ and $gl\,(n, C)$ be the space of all real and complex n-*by*-n matrices. These are the linear Lie algebra of $GL(n, R)$ and $GL(n, C)$ respectively. For some particular other closed linear group we can compute the linear Lie algebra explicitly.

For example, with $G = SO(3)$, we have alredy used a curve involving cosines and sines to see that the matrix $\begin{pmatrix} 0 & 1 & 0 \\ -1 & 0 & 0 \\ 0 & 0 & 0 \end{pmatrix}$ is one member of the linear Lie algebra $\mathfrak{g}$ of $SO(3)$. Using different curves involving cosines and sines, we can see that the matrices

$\begin{pmatrix} 0 & 0 & 1 \\ 0 & 0 & 0 \\ -1 & 0 & 0 \end{pmatrix}$ and $\begin{pmatrix} 0 & 0 & 0 \\ 0 & 0 & 1 \\ 0 & -1 & 0 \end{pmatrix}$ lie in $\mathfrak{g}$ also. Taking into account the vector space operations, we see that

$$\mathfrak{g} \supseteq = \left\{ \begin{pmatrix} 0 & a & b \\ -a & 0 & c \\ -b & -c & 0 \end{pmatrix} \right\} = \left\{ \begin{pmatrix} 0 & a & b \\ -a & 0 & c \\ -b & -c & 0 \end{pmatrix} \right\} = \{\text{skew-symmetric real } 3\text{-}by\text{-}3 \text{ matrices}\}.$$

Actually equality holds in this inclusion. In fact, if $c(t)(c(t))' = 1$, then

$$\begin{aligned} \dot{c}\,(t)\,(c(t))' + c(t)\,(\dot{c}\,(t))' &= 0 \\ \dot{c}\,(0)\,1' + 1\ \dot{c}\,(0)' &= 0 \\ \dot{c}\,(0) + \dot{c}\,(0)' &= 0 \qquad (13.13.7) \end{aligned}$$

and $\dot{c}\,(0)$ is skew symmetric. Thus the linear Lie algebra of $SO(3)$ is

$$\begin{aligned} so(3) &= \{\text{skew-symmetric real } 3\text{-}by\text{-}3 \text{ matrices}\} \\ &= \{X \in gl\,(3, R) \mid X + X' = 0\} \end{aligned}$$

For the five examples in (13.13.1), we define

$$so(n) = \{X \in gl\,(n, R) \mid X + X' = 0\}$$

$$u(n) = \{X \in gl\,(n, C) \mid X + X^* = 0\}$$
$$su(n) = \{X \in gl\,(n, C) \mid X + X^* = 0 \text{ and } \mathrm{Tr}\, X = 0\}$$
$$sl\,(n, R) = \{X \in gl\,(n, R) \mid \mathrm{Tr}\, X = 0\}$$
$$sl\,(n, C) = \{X \in gl\,(n, C) \mid \mathrm{Tr}\, X = 0\} \qquad (13.13.8)$$

these are the respective linear Lie algebra of the closed linear groups in (13.13.1).

(13.14) *Definiton*:

Let g be a Lie algebra over A, let g be a subset of g form the basis of g considered as a vector space.

A representation of g on V is a homomorphism ρ: g $\rightarrow \Im(V)$. That is, ρ satisfies the conditions:

(1) $\rho(\alpha) \in \Im(V)$ for all $\alpha \in$ g.

(2) $\rho([\alpha, \beta]) = [\rho(\alpha), \rho(\beta)]$

(3) $\rho\,(a\,\alpha + b\,\beta) = a\,\rho(\alpha) + b\,\rho(\beta)$ a ,b $\in$ A, α, $\beta \in$ g

A subspace W of V is said to be invariant under ρ if $\rho(\alpha)w \in$ W for all $\alpha \in$ g, w $\in$ W . A representation ρ of g on V is irreducible if there is no proper subspace W of V which is invariant under ρ.

Two vector space V, V′ over A are isomorphic if there exists a mapping μ of V into V′

(1) $\mu(a_1 v_1 + a_2 v_2) = a_1\,\mu(v_1) + a_2\,\mu(v_2)$.

(2) If $\mu(v) = 0'$ then v = 0.

(3) For every v′ $\in$ V′ there exists a $v \in$ V such that $\mu(v) = v'$.

14

Covariant Differentiation

(14.1) Suppose m be a C^k morphism of a C^k manifold Z into a C^k manifold X, let (η, Y, X) be a C^k vector bundle over X. Then a C^k morphism g of Z into Y is called a lifting of m to η iff $\eta \circ g = m$ we denote by $Lift_{\eta}^{k}(m)$ the set of all lifting of m to η. If $Z = X$, then

$Lift_{\tau_X}^{k}(id_X) = \Gamma^k(X)$ = set of c^k vector fields on X.

Any C^k vector field w on Z is a lifting of id_Z to τ_Z and for any vector field v on X, $v \circ m$ is a C^k lifting of m to τ_X. Also since $\tau_X \circ Tm \circ w = m \circ \tau_Z \circ w$, it follows that $Tm \circ w$ is a lifting of m to τ_X.

Let ∇^z be a function on $T_z Z \times Lift^k(m)$ into $\eta^{-1}[m(z)]$ where z is any point of Z, then ∇^z is called a covariant differentiation at z relative to m iff the following conditions are satisfied

(CD: 1) For each pair of vectors v and w in $T_z Z$ and each lifting f in $Lift^k(m)$

$$\nabla^Z_{v+w}(f) = \nabla^Z_v(f) + \nabla^Z_w(f)$$

(CD: 2) For each vector v in $T_z Z$ and each lifting g and h in $Lift^k(m)$

$$\nabla^Z_v(f+g) = \nabla^Z_v(f) + \nabla^Z_v(g)$$

(CD: 3) For each vector v in $T_z Z$ and every scalar a in A, and each lift f in $Lift^k(m)$

$$\nabla^Z_{av}(f) = a\nabla^Z_v(f)$$

(CD: 4) For each vector v in $T_z Z$ and each lifting f in $Lift^k(m)$ and a C^k morphism h of Z into A.

$$\nabla^Z_v (hf) = h(z)\, \nabla^Z_v (f) + ((d_z h)(v))\, f(z)$$

The function ∇ on $\Gamma^k(Z) \times Lift^k(m)$ into $Lift^k(m)$ such that for each vector field v in $\Gamma^k(z)$ and each f in $Lift^k(m)$, $\nabla_v f$ is in $Lift^k(m)$ and so for all z in Z, $(\nabla_v f)(z) = \nabla^Z_{v(z)} f$ called a covariant differentiation relative to m.

A covariant differentiation in X is a covariant differentiation relative to id_X such that for pair of vector fields v and w on X, $\nabla_v w$ is also a vector field on X and for each x in X, $(\nabla_v w)(x) = \nabla^x_{v(x)} w$.

Let us suppose that X be a manifold modelled on A^n. Let (U, θ) be a chart for X, and $(g_i, i \in \underline{n})$ be a list of basic vector fields on U. Let ∇ be a covariant differentiation in X, define a function Γ^i_{jk} on U into A where $(i, j, k) \in \underline{n} \times \underline{n} \times \underline{n}$. Then for each x in U

$$\Gamma^i_{jk}(x) = (d_x \theta^i)((\nabla_{gj} . g_k)(x))$$

such a function Γ^i_{jk} is called a component of ∇, evidently

$$(\nabla_{g_j} \cdot g_k)(x) = \sum_{i \in \underline{n}} \Gamma^i_{jk}(x)(g_i(x))$$

suppose v be any vector in $T_x X$ and w be any vector field on X.

$$\text{Then } v = \sum_{i \in \underline{n}} a_i g_i(x) \text{ and } w = \sum_{i \in \underline{n}} b_i g_i$$

where $(b_i, i \in \underline{n})$ be a list of C^k morphism of U into A and $(a_i, i \in \underline{n})$ be a list of scalar in A. Now

$$\nabla^X_v(w) = \nabla^X_{\sum\limits_{i \in \underline{n}} a_i g_i(x)} \sum_{j \in \underline{n}} b_j g_j$$

$$= \sum_{i \in \underline{n}} a_i \nabla^X_{g_i(x)} \sum_{j \in \underline{n}} b_j g_j$$

$$= \sum_{(i,j) \in \underline{n} \times \underline{n}} a_i \nabla^X_{g_i(x)} \cdot b_j g_j$$

$$= \sum_{(i,j) \in \underline{n} \times \underline{n}} \Big(a_i \big(b_j(x)\, \nabla^X_{g_i(x)}\, g_j + \big((d_x b_j)(g_i(x))\big) g_j(x) \big) \Big)$$

$$= \sum_{(i,j)\in \underline{n}\times\underline{n}} \left(a_i\left(b_j(x)\left(\nabla_{g_i} g_j\right)(x) + \left(\left(d_x b_j\right)\left(g_i(x)\right)\right)g_j(x)\right)\right)$$

$$= \sum_{(i,j)\in \underline{n}\times\underline{n}} \left(a_i\left(b_j(x)\sum_{k\in\underline{n}}\Gamma_{ij}^k(x)\left(g_k(x)\right)\right) + \left(\left(d_x b_j\right)\left(g_i(x)\right)\right)g_j(x)\right)$$

$$= \sum_{(i,j)\in \underline{n}\times\underline{n}} \left(a_i\left(\left(d_x b_j\right)\left(g_i(x)\right)\right)g_j(x) + a_i\, b_j(x)\sum_{k\in\underline{n}}\Gamma_{ij}^k(x)\left(g_k(x)\right)\right)$$

changing the indices

$$= \sum_{(j,k)\in \underline{n}\times\underline{n}} \left(a_j\left(\left(d_x b_k\right)\left(g_j(x)\right)\right)g_k(x) + a_j\, b_k(x)\sum_{i\in\underline{n}}\Gamma_{jk}^i(x)\left(g_i(x)\right)\right)$$

(14.2) Suppose X and Z be an open subsets of Banach space R and Q over A respectively. Let m be a C^kmorphism of Z into X and let $(pr_1, X\times S, X)$ be a trivial vector bundle over X of fiber type a Banach space S over A. Then each *lifting* of m to pr_1 is evidently of the form $f = (m, f_2)$, where f_2 is a C^k morphism of Z into S. Let B be a C^∞ morphism of $X\times R\times S$ into S such that for each x in X, $(r, s) \mapsto B(x, r, s)$ is a bilinear map of $R\times S$ into S.

Let z be any point of Z, then the function ∇^z on $(\{z\}\times Q)\times Lift^k(m)$ into $\{m(z)\}\times S$ such that for each (z, q) in $\{z\}\times Q$ and each $f = (m, f_2)$ in $Lift^k_{pr_1}(m)$

$$\nabla^z_{(z,q)}(f) = (m(z), (Df_2(z)(q) + B(m(z), Dm(z)(q), f_2(z))))$$

is a covariant differentiation at z relative to m.

Proof: To prove ∇^z is a covariant differentiation at z relative to m it is sufficent to satisfy (C.D.1) through (C.D.4), we have

(C D.1) $\nabla^z_{(z,q_1)+(z,q_2)}(f) = \nabla^z_{(z,q_1+q_2)}(f)$

$$= (m(z), (Df_2(z))(q_1+q_2) + B(m(z), Dm(z)(q_1+q_2), f_2(z)))$$

$$= (m(z), (Df_2(z).q_1+Df_2(z).q_2 + B(m(z), Dm(z).q_1, f_2(z)) + B(m(z), Dm(z).q_2, f_2(z))))$$

$$= (m(z), (Df_2(z).q_1 + B(m(z), Dm(z).q_1, f_2(z)) + Df_2(z).q_2 + B(m(z), Dm(z).q_2, f_2(z))))$$

$$= (m(z), (Df_2(z).q_1 + B(m(z), Dm(z).q_1, f_2(z))) + (m(z), (Df_2(z).q_2 + B(m(z), Dm(z).q_2, f_2(z)))$$

$$= \nabla^z_{(z,q_1)}(f) + \nabla^z_{(z,q_2)}(f)$$

(C D.2) $\nabla^z_{(z,q)}(f+g)$

$$= (m(z), (D(f_2 + g_2)(z)(q) + B(m(z), Dm(z)(q), (f_2 + g_2)(z))))$$

$$= (m(z), (D f_2 (z).q + D g_2 (z).q + B(m(z), Dm(z).q, f_2(z)) + B(m(z), Dm(z).q, g_2(z)))$$

$$= (m(z), (D f_2 (z).q + B(m(z), Dm(z).q, f_2(z))) + (m(z), (D g_2 (z).q + B(m(z), Dm(z).q, g_2(z)))$$

$$= \nabla^z_{(z,q)}(f) + \nabla^z_{(z,q)}(g)$$

(C D.3) $\nabla^z_{a(z,q)}(f) = \nabla^z_{(z,aq)}(f)$

$$= (m(z), (Df_2(z).\ aq + B(m(z), Dm(z).aq, f_2(z)))$$

$$= (m(z), a\,(Df_2(z).\ q + B(m(z), Dm(z).q, f_2(z)))$$

$$= a\,(m(z), (Df_2(z).\ q + B(m(z), Dm(z).q, f_2(z)))$$

$$= a\nabla^z_{(z,q)}(f)$$

(C D.4) $\nabla^z_{(z,q)}(hf)$

$$= (m(z), (D(hf_2)(z).\ q + B(m(z), Dm(z).q, hf_2)(z))))$$

$$= (m(z), ((Dh(z).\ q\ f_2(z) + Df_2(z).\ q\ \ h(z)) + B(m(z), Dm(z).q, (f_2(z).h(z))))$$

$$= (m(z), (Dh(z).\ q\ ) f_2(z) + h(z)\,(Df_2(z).\ q + B(m(z), Dm(z).q, f_2(z)))$$

$$= (m(z), (Dh(z).\ q\ ) f_2(z)) + (m(z), h(z).(Df_2(z).\ q + B(m(z), Dm(z).q, f_2(z)))$$

$$= Dh(z).\ q\ .(m(z), f_2(z)) + h(z)\ .\ (m(z), (Df_2(z).\ q + B(m(z), Dm(z).q, f_2(z)))$$

$$= Dh(z).\ q\ .\ f(z) + h(z)\ \nabla^z_{(z,q)}(f)$$

(14.3) Let X be a C^∞ manifold modelled on A^n. Let (U, θ) be a chart for X, Let ∇ be a covariant differentiation in U, let $\{e_i; i \in \underline{n}\}$

be the list of the basic vector fields on U relative to θ and let x be any point of U. Let u, v, w are vector fidds on $U \subset X$ and p, q, r are the vectors in $T_x U$. Then

$$p = \sum_{i \in \underline{n}} a_i e_i(x),$$

$$q = \sum_{i \in \underline{n}} b_i e_i(x),$$

$$r = \sum_{i \in \underline{n}} c_i e_i(x)$$

where $\{a_i;\ i \in \underline{n}\}, \{b_i;\ i \in \underline{n}\}$ and $\{c_i;\ i \in \underline{n}\}$ are list of elements of A.

$$\text{and } u = \sum_{i \in \underline{n}} f_i e_i,$$

$$v = \sum_{i \in \underline{n}} g_i e_i,$$

$$w = \sum_{i \in \underline{n}} h_i e_i,$$

where $\{f_i;\ i \in \underline{n}\}, \{g_i;\ i \in \underline{n}\}$ and $\{h_i;\ i \in \underline{n}\}$ are list of C^∞ morphism of U into A. Let $\{t_i;\ i \in \underline{2n}\}$ be the list of the basic vector fields on TU relative to θ. Then by (5.5) and (8.5)

$$(T_x u)(p) = \sum_{i \in \underline{n}} \Big(\big(d_x (\underset{\sim}{\theta}^i \circ u)\big)(p) \Big) t_i(u(x)) + \sum_{i \in \underline{n}} \Big(\big(d_x (\underset{\sim}{\theta}^{n\cdot i} \circ u)\big)(p) \Big) t_{n\cdot i}(u(x))$$

$$= \sum_{i \in \underline{n}} \big((d_x \theta^i)(p) \big) t_i(u(x)) + \sum_{i \in \underline{n}} \Big(\big(d_x (\underset{\sim}{\theta}^{n\cdot i} \circ u)\big)(p) \Big) t_{n\cdot i}(u(x))$$

$$= \sum_{i \in \underline{n}} \left((d_x \theta^i) \left(\sum_{i \in \underline{n}} a_i e_i(x) \right) t_i(u(x)) \right)$$

$$+ \sum_{i \in \underline{n}} \left(\big(d_x (\underset{\sim}{\theta}^{n\cdot i} \circ u)\big) \left(\sum_{j \in \underline{n}} a_j e_j(x) \right) \right) t_{n\cdot i}(u(x))$$

$$= \sum_{i \in \underline{n}} a_i t_i(u(x)) + \left(\sum_{(i,j)\, \underline{n} \in \underline{n}} a_i \left(\left(d_x \left(\underset{\sim}{\theta}^{n \cdot i} \circ u \right) \right) e_j(x) \right) \right) t_{n \cdot i}(u(x))$$

$$= \sum_{i \in \underline{n}} a_i t_i(u(x)) + \left(\sum_{(i,j)\, \underline{n} \in \underline{n}} a_i ((d_x f) e_i(x)) \right) t_{n \cdot i}(u(x))$$

$$X \xrightarrow{\ \bar{z}\ } T_x X,\ U \xrightarrow{\ u\ } TU \xrightarrow{\ \theta^{n \cdot i}\ } A$$

$$T_x X \to T(TX)$$

consequently, for each point $y \in T_x X$ the set of vectors of the form $(T_x z)(y)$, where z is a vector field on X such that $z(x) = s$ and y is a points of $T_x X$, generates $T(T_x X)$

If $(T_x u)(p) = (T_x v)(q)$, then since $(T_x u)(p)$ is in $T_{u(x)}(TU)$ and $(T_x v)(q)$ is in $T_{v(x)}(TU)$ implies $u(x) = v(x)$

now $\quad u(x) = \sum_{i \in \underline{n}} f_i(x) e_i(x)$

and $\quad v(x) = \sum_{i \in \underline{n}} g_i(x) e_i(x)$

so $\quad u(x) = v(x) \Rightarrow \sum_{i \in \underline{n}} (f_i(x) - g_i(x)) e_i(x) = 0$

since $e_i(x)$ are linearly independent vectors in TU, $f_i(x) = g_i(x)$

Now $(T_x u)(p) = (T_x v)(q)$

$$\Rightarrow \sum_{i \in \underline{n}} a_i t_i(u(x)) + \left(\sum_{(i,j)\, \underline{n} \in \underline{n}} a_j ((d_x f_i) e_j(x)) \right) t_{n \cdot i}(u(x))$$

$$= \sum_{i \in \underline{n}} b_i t_i(v(x)) + \left(\sum_{(i,j)\, \underline{n} \in \underline{n}} b_j ((d_x g_i) e_j(x)) \right) t_{n \cdot i}(v(x))$$

so using u(x) = v(x) and $f_i(x) = g_i(x)$

we have $a_i = b_i$. Hence $p = q$.

Again if $(T_x u)(p) = (T_x v)(q)$, then writing the expression for $\nabla^x_p(u)$ and $\nabla^x_q(v)$ we get

$$\nabla^x_p(u) = \sum_{(j,k)\in \underline{n}\times\underline{n}} a_j((d_x f_k)e_j(x))e_k(x) + a_j f_k(x)\left(\sum_{i\in\underline{n}} \Gamma^i_{jk}(x)\, e_i(x)\right)$$

$$\nabla^x_q(v) = \sum_{(j,k)\in \underline{n}\times\underline{n}} b_j((d_x g_k)e_j(x))e_k(x) + b_j g_k(x)\left(\sum_{i\in\underline{n}} \Gamma^i_{jk}(x) e_i(x)\right)$$

Since $(T_x u)(p) = (T_x v)(q)$

$\Rightarrow a_i = b_i$ and $f_i(x) = g_i(x)$ for all i

$\Rightarrow \nabla^x_p(u) = \nabla^x_q(v)$

If $(T_x w)(r) = (T_x u)(p) + (T_x v)(q)$, then $w(x) = u(x) - v(x)$, because $(T_x w)(r)$ is in $T_{w(x)}TU$ and $(T_x u)(p) = (T_x v)(q)$. Keeping in view the above observation, for a covariant differentiation in X we define a function K_∇ on $T(T(X))$ into $T(X)$, calling it the connector of ∇, such that for ecah x in X and each p in $T_x X$ and each vector field u in $\Gamma^\infty(X)$

$$K_\nabla((T_x u)(p)) = \nabla^x_p(u)$$

we note that for each x in X, each p in $T_x X$ and each u in $\Gamma^\infty(X)$ the restriction of K_∇ to $T_{u(x)}(T(X))$ is a linear map of $T_p(T(X))$ into $T_x X$

(14.4) let X be a C^∞ manifold modelled on A^n, let (U, θ) be a chart for X, let ∇ be a covariant differentiation in U, let $(e_i; i \in \underline{n})$ be the list of the basic vector fields on U relative to θ and let $(t_i; i \in \underline{2n})$ be the list of the basic vector fields on TU, relative to $\underset{\sim}{\theta}$. Let x be any points of U and let p be any vector in $T_x U$. Let F be a function on $T_p(TU)$ into $T_x U$ such that for each z in $T_p(TU)$

$$F(z) = \sum_{i\in\underline{n}} (d_p \underset{\sim}{\theta}^{n+i})(z)e_i(x)$$
$$+ \sum_{(i,j,k)\in\underline{n}\times\underline{n}\times\underline{n}} (d_p \underset{\sim}{\theta}^j)(z)(d_x\theta^k)(p)\Gamma^i_{jk}(x)e_i(x)$$

Then F is identical with the restriction of K_∇ to $T_p(TU)$.

Proof: Let v be any C^∞ vector field on U such that $v(x) = p$ and let q be any vector in $T_x U$. Then to show the identity of F with the restrication of K_∇ to $T_p(TU)$ it suffices to show that for every vector of the form $(T_x v)(q)$ in $T_{v(x)}(TU) = T_p(TU)$

$$F((T_x v)(q)) = K_\nabla((T_x v)(q)) = \nabla_q^x(v)$$

Put $q = \sum_{i \in \underline{n}} b_i e_i(x)$ and

$$v = \sum_{i \in \underline{n}} g_i e_i$$

where $(b_i ; i \in n)$ is a list of elements of A and $(g_i ; i \in n)$ is a list of C^∞ morphism of U into A. Then by (14.4), since $v(x) = p$

$$(T_x v)(q) = \sum_{r \in \underline{n}} b_r t_r(p) + \sum_{(r,s) \in \underline{n} \times \underline{n}} b_s(d_x g_r)(e_s(x)) t_{n+r}(p)$$

Now

$$F((T_x v)(q)) = \sum_{i \in \underline{n}} (d_p \underset{\sim}{\theta}^{n+i})((T_x v)(q)) e_i(x)$$

$$+ \sum_{(i,j,k) \in \underline{n} \times \underline{n} \times \underline{n}} (d_p \underset{\sim}{\theta}^j)((T_x v)(q))(d_x \theta^k)(p) \Gamma_{jk}^i(x) e_i(x) \tag{14.3.1}$$

Hence first taking the first part of (14.3.1) we have

$$\sum_{i \in \underline{n}} (d_p \underset{\sim}{\theta}^{n+i})((T_x v)(q)) e_i(x)$$

$$= \sum_{i \in \underline{n}} (d_p \underset{\sim}{\theta}^{n+i})\Big(\sum_{r \in \underline{n}} b_r t_r(p)$$

$$+ \sum_{(r,s) \in \underline{n} \times \underline{n}} b_s(d_x g_r)(e_s(x)) t_{n+r}(p)\Big) e_i(x)$$

$$= \sum_{(i,s) \in \underline{n} \times \underline{n}} b_s(d_x g_i)(e_s(x)) e_i(x)$$

and taking the second part (14.3.1) we have

$$\sum_{(i,j,k) \in \underline{n} \times \underline{n} \times \underline{n}} (d_p \underset{\sim}{\theta}^j)((T_x v)(q))(d_x \theta^k)(p) \Gamma_{jk}^i(x) e_i(x)$$

$$= \sum_{(i,j,k)\,\in\,\underline{n}\times\underline{n}\times\underline{n}} (d_p \underset{\sim}{\theta}^j)\big(\sum_{r\in\underline{n}} b_r t_r(p)$$

$$+ \sum_{(r,s)\,\in\,\underline{n}\times\underline{n}} b_s(d_x g_r)(e_s(x))t_{n+r}(p))(d_x\theta^k)(p)\Gamma^i_{jk}(x)e_i(x)$$

$$= \sum_{(i,j,k)\,\in\,\underline{n}\times\underline{n}\times\underline{n}} b_j((d_x\theta^k)(p))\Gamma^i_{jk}(x)e_i(x)$$

Thus we have

$$F((T_x\ v)(q)) = \sum_{(i,s)\,\in\,\underline{n}\times\underline{n}} b_s(d_x g_i)(e_s(x))e_i(x)$$

$$+ \sum_{(i,j,k)\,\in\,\underline{n}\times\underline{n}\times\underline{n}} b_j((d_x\theta^k)(p))\Gamma^i_{jk}(x)e_i(x)$$

Now

$$\nabla^x_q(v) = \nabla^x_{\sum_{i\in\underline{n}} b_i e_i(x)}\big(\sum_{j\in\underline{n}} g_j e_j\big)$$

$$= \sum_{j\in\underline{n}} \nabla^x_{\sum_{i\in\underline{n}} b_i e_i(x)}(g_j e_j)$$

$$= \sum_{(i,j)\,\in\,\underline{n}\times\underline{n}} \nabla^x_{b_i e_i(x)} g_j e_j$$

$$= \sum_{(i,j)\,\in\,\underline{n}\times\underline{n}} b_i \nabla^x_{e_i(x)} g_j e_j$$

$$= \sum_{(i,j)\,\in\,\underline{n}\times\underline{n}} b_i(g_j\nabla^x_{e_i(x)}(e_j) + (d_x g_j)(e_i(x))e_j(x))$$

$$= \sum_{(i,j)\,\in\,\underline{n}\times\underline{n}} (b_i g_j \sum_{k\in\underline{n}} G^k_{ij}(x)\,e_k(x) + b_i(d_x g_j)(e_i(x))\,e_j(x))$$

$$= \sum_{(i,j,k)\,\in\,\underline{n}\times\underline{n}\times\underline{n}} b_i g_j G^k_{ij}(x)\,e_k(x) + \sum_{(i,j)\,\in\,\underline{n}\times\underline{n}} b_i(d_x g_j)(e_i(x))\,e_j(x)$$

Hence $F((T_x\ v)(q)) = K_\nabla((T_x\ v)(q)) = \nabla^x_q(v)$

Thus for each z in $T_p(TU)$

$$K_\nabla(z) = \sum_{i \in \underline{n}} (d_p \underset{\sim}{\theta}^{n+i})(z) e_i(x)$$

$$+ \sum_{(i,j,k) \in \underline{n} \times \underline{n} \times \underline{n}} (d_p \underset{\sim}{\theta}^j)(z)(d_x \theta^k)(p) \Gamma^i_{jk}(x) e_i(x)$$

(14.5) Let X be a C^∞ manifold modelled on A^n and let ∇ be a covariant differentiation in X. Let a be any element of A and let h_a be the function on TX into itself such that for each v in TX, $h_a(v) = av$. Then

$$K_\nabla \circ (Tha) = h_a \circ K_\nabla$$

Proof: Let w be any elememt of TX. Then the restriction of K_∇ and h_a to $T_w(TX)$ are linear, and so it suffices to show that for each x in X, each v in $T_x X$ and each vector field z in $\Gamma^\infty(X)$

$$K_\nabla((Th_a))((Tz)(v)) = h_a \circ K_\nabla((Tz)(v))$$

Now

$$\begin{aligned} K_\nabla((Th_a))((T_x z)(v)) &= K_\nabla((T_x(h_a \circ z))(v)) \\ &= K_\nabla((T_x(a\, z))(v)) \\ &= \nabla^x_v(a\, z) \\ &= a\nabla^x_v(z) \\ &= h_a(K_\nabla((T_x z)(v))) \end{aligned}$$

(14.6) Let X be a C^∞ manifold modelled on A^n and let ∇ be a covariant differentiation in X. Then any element z of $T(T(x))$ is said to be vertical (respectively horizontal), if $T(\tau_X)(z) = 0$ (respectively $K_\nabla(z) = 0$). If an element z of $T(T(X))$ is both vertical and horizontal then , by (9.1) and (14.4), $z = 0$ hence for each v in TX the map

$$T_v(TX) \ni z \to (T(\tau_X)(z), K_\nabla(z)) \in T_{\tau_X(v)}(X) \times T_{\tau_X(v)}(X)$$

is a linear isomorphism, and so $T_v(TX)$ is the direct sum of its vector subspace consisting of the vertical and horizontal vectors in $T_v(TX)$.

(14.7) Let m be a C^∞ morphism of a C^∞ manifold Z modelled on A^p into a C^∞ manifold X modelled on A^n. Let ∇ be a covariant differentiation in X. For each z in Z, each v in $T_z Z$ and each f in $Lift^k(m)$ put

$$\overline{\nabla}^z_v(f) = K_\nabla((T_z f)(v))$$

Then for each z in Z, $\overline{\nabla}^z$ is a covariant differentiation at z relative to m.

Proof: Let z be any point of Z, then $\overline{\nabla}^z$ clearly satisfies (C D.1) and (C D.3), To verify (C D.2) and (C D.4) let (U,θ) be a chart at $m(z)$ in X. Let $(e_i; i \in \underline{n})$ be the list of the basic vector fields on U relative to θ, and let $(t_i; i \in \underline{2n})$ be the list of the basic vector field on TU relative to $\underset{\sim}{\theta}$. Then by (5.5) each v in $T_z Z$ and each f in $Lift^k(m)$, $(Tf)(v)$ is in $T_v(TX)$ hence by (5.5),

$$z \in m^{-1}[U] \Rightarrow f(z) = \sum_{i \in \underline{n}} (d_{m(z)}\theta^i)(f(z))e_i(m(z))$$

By (5.5) for each v in $T_z Z$ and each f in $lift^\infty(m)$

$$(T_z f)(v) = \sum_{i \in \underline{2n}} d_z(\underset{\sim}{\theta}^i \circ f)(v)t_i(f(z))$$

hence by (5.5), (8.5) and (14.4)

$$\begin{aligned}\overline{\nabla}^z_v(f) &= K_\nabla((Tf)(v)) \\ &= \sum_{i \in \underline{n}} (d_z(\underset{\sim}{\theta}^{n+i} \circ f)(v)e_i(m(z)) \\ &+ \sum_{(i,j,k) \in \underline{n} \times \underline{n} \times \underline{n}} (d_z(\underset{\sim}{\theta}^j \circ m)(v)(d_{m(z)}q^k)(f(z))G^i_{jk}(m(z))e_i(m(z)\end{aligned}$$

In particular for each v in $T_z Z$ and each r in $\underline{n}$

$$\overline{\nabla}^z_v(e_r \circ m) = \sum_{(i,j) \in \underline{n} \times \underline{n}} (d_z(\underset{\sim}{\theta}^j \circ m)(v)G^i_{jr}(m(z))e_i(m(z))$$

Consequently for each v in $T_z Z$ and each f in $Lift^k(m)$

$$\overline{\nabla}^z_v(f) = \sum_{k \in \underline{n}} (d_z(\underset{\sim}{\theta}^{n+k} \circ f)(v)e_k(m(z)) + ((d_{m(z)}\theta^k)(f(z))\overline{\nabla}^z_v(e_k \circ m)$$

For each v in $T_z Z$, each pair of elements g and h in $Lift^k(m)$ and each k in $\underline{n}$

$$(d_z(\underset{\sim}{\theta}^{n-k} \circ (g+h))(v)) = (d_z(\underset{\sim}{\theta}^{n-k} \circ g)(v)) + (d_z(\underset{\sim}{\theta}^{n-k} \circ h)(v))$$

and

$$(d_{m(z)}\theta^k)((g+h)(z)) = (d_{m(z)}\theta^k)(g(z)) + (d_{m(z)}\theta^k)(h(z))$$

hence

$$\overline{\nabla}^z_v(g+h) = \overline{\nabla}^z_v(g) + \overline{\nabla}^z_v(h)$$

Thus (C. D. 2) is verified.

For each v in $T_z Z$ and each f in $Lift^\infty(m)$, each C^k morphism h of $m^{-1}[U]$ into A, and each k in $\underline{n}$

$$(d_z(\underset{\sim}{\theta}^{n-k} \circ (hf))(v)) = ((d_z h)(v))(d_{m(z)}\theta^k)(f(z)) + h(z)(d_z(\underset{\sim}{\theta}^{n-k} \circ f)(v))$$

hence

$$\begin{aligned}\overline{\nabla}^z_v(hf) &= \sum_{k\in\underline{n}}(d_z \underset{\sim}{\theta}^{n-k} \circ (hf))(v)e_k(m(z)) + ((d_{m(z)}\theta^k)(h(z)f(z))\overline{\nabla}^z_v(e_i \circ m) \\ &= \sum_{k\in\underline{n}}((d_z h)(v))(d_{m(z)}\theta^k)(f(z))e_k(m(z)) + h(z)(d_z(\underset{\sim}{\theta}^{n-k} \circ f)(v))e_k(m(z)) + h(z)((d_{m(z)}\theta^k)(f(z))\overline{\nabla}^z_v(e_k \circ m) \\ &= \sum_{k\in\underline{n}}(((d_z h)(v))(d_{m(z)}\theta^k)(f(z)))e_k(m(z)) + h(z)\,\overline{\nabla}^z_v(f) \\ &= \sum_{k\in\underline{n}}((d_z h)(v))(f(z)) + h(z)\,\overline{\nabla}^z_v(f)\end{aligned}$$

thus (C D. 4) is verified.

(14.8) Let m be a C^∞ morphism of a C^∞ manifold Z modelled on A^p into a C^∞ manifold X modelled on A^n and let ∇ be a covariant differentiation in X, then the covariant differentiation relative to m which assigns to each z in Z, each w in $\Gamma^\infty(Z)$ and each f in $Lift^\infty(m)$ the value

$$K_\nabla((Tf)(w(z)))$$

is called the pull back of ∇ by m and is denoted by $m^*\nabla$. Let $(e_i;\ i \in \underline{n})$ be the list of the n basic vector fields relative to a chart θ for X. Let f be any member of $Lift^\infty(m)$, and let w be any member of $\Gamma^\infty(Z)$. Then for each z in Z

$$f(z) = \sum_{k \in \underline{n}} (d_{m(z)}\theta^k)(f(z))e_k(m(z))$$

and

$$\begin{aligned}((m^*\nabla)_w f)(z) &= (m^*\nabla)^Z_{w(z)} f \\ &= K_\nabla((Tf)(w(z))) \\ &= \sum_{k \in \underline{n}} (d_z(\underset{\sim}{\theta}^{n\cdot k} \circ f)(w(z))e_k(m(z)) \\ &\qquad + ((d_{m(z)}\theta^k)(f(z))(m*\nabla)^z_w(e_k \circ m) \\ &= \sum_{k \in \underline{n}} ((d_z f_k)(w(z)))e_k(m(z)) \\ &\qquad + (f_k(z))(m*\nabla)^z_w(e_k \circ m)\end{aligned}$$

where f_k is the function $Z \ni z \mapsto (d_{m(z)}\theta^k)(f(z)) \in A$ for each k in $\underline{n}$. For each k in $\underline{n}$ and each z in Z by (14.6) and (14.2)

$$(m*\nabla)^z_w(e_k \circ m) = \sum_{j \in \underline{n}} ((d_z(\theta^j \circ m))(w))(\nabla_{e_j} e_k)(m(z)$$

Hence for each z in Z

$$\begin{aligned}(m*\nabla)_w f)(z) &= \sum_{k \in \underline{n}} (((d_z f_k)(w(z)))e_k(m(z)) \\ &+ \sum_{(j,k) \in \underline{n} \times \underline{n}} f_k(z)((d_z(\theta^j \circ m))(w))\nabla_{e_j} e_k)(m(z))\end{aligned}$$

(14.9) Let m be C^∞ morphism of a C^∞ manifold Z modelled on A^p into a C^∞ manifold X modelled on A^n, and let ∇ be a covariant differentiation in X. Then for each z in Z, each w in $T_z Z$ and each

$$v = \sum_{k \in \underline{n}} h_k e_k$$

where $(h_k;\, k \in \underline{n})$ is a list of C^k morphism of $Dm(\theta)$ into A and $(e_i;\, i \in \underline{n})$ is the list of the basic vector fields on $Dm(\theta)$ relative to a chart θ at $m(z)$.

$$(m * \nabla)^z_w(v \circ m) = \sum_{k \in \underline{n}} (m * \nabla)^z_w (h_k \circ m)(e_k \circ m))$$

$$= \sum_{k \in \underline{n}} (d_z(h_k \circ m)(w))(e_k(m(z)) + \sum_{k \in \underline{n}} (h_k(m(z)))(m * \nabla)^z_w(e_k \circ m)$$

$$= \sum_{k \in \underline{n}} (d_{m(z)} h_k)((T_z m)(w))(e_k(m(z))$$

$$+ \sum_{(i,j,k) \in \underline{n} \times \underline{n} \times \underline{n}} h_k(m(z))(d_z(\theta^j \circ m)(w))\Gamma^i_{jk}(m(z))e_i(m(z))$$

(by (5.3), (5.4) and (14.7))

and in view of the fact that

$$(Tm)(w) = \sum_{j \in \underline{n}} (d_z(\theta^j \circ m)(w))e_j(m(z)$$

by (9.1)

$$\nabla^{m(z)}_{(Tm)(w)}(v) = \sum_{(j,k) \in \underline{n} \times \underline{n}} (d_z(\theta^j \circ m)(w))(d_{m(z)} h_k)(e_j(m(z))e_k(m(z))$$

$$+ \sum_{(i,j,k) \in \underline{n} \times \underline{n} \times \underline{n}} h_k(m(z))(d_z(\theta^j \circ m)(w)\Gamma^i_{jk}(m(z))e_i(m(z))$$

by (14.1)

$$= \sum_{k \in \underline{n}} (d_{m(z)} h_k)((T_z m)(w))(e_k(m(z))$$

$$+ \sum_{(i,j,k) \in \underline{n} \times \underline{n} \times \underline{n}} h_k(m(z))(d_z(\theta^j \circ m)(w)\Gamma^i_{jk}(m(z))e_i(m(z))$$

$$= (m * \nabla)^z_w(v \circ m)$$

(14.10) Let X be a separated smooth manifold modelled on $\underline{R}^n$ with covariant diefferetiation ∇, Let c be a smooth map of closed interval I of $\underline{R}$ into X, let f be a map of I into TX such that $\tau_x \circ f = c$, let t be any element of I, and let v and w be any element of $T_{c(t)}X$. Then there exists a unique smooth map g of I into TX such that $\tau_x \circ g = c$,

$$(c^*\nabla)_{\bar{1}}((c^*\nabla)_{\bar{1}}g)+f=0,\ g(t)=v,\ ((c^*\nabla)_1 g)(t)=w.$$

Proof: without loss of generality we suppose that there exists a chart θ for X such that $c(I)\subset Dm(\theta)$. Consider a smooth map g of I into TX such that $\tau_x \circ g = c$. Let $(e_i; i \in \underline{n})$ be the list of the n basic vector fields relative to θ. Then by (14.9)

$$f=\sum_{i\in\underline{n}} f_i(e_i \circ c),\quad g=\sum_{i\in\underline{n}} g_i(e_i \circ c),$$

where for each i in $\underline{n}$. f_i and g_i are smooth maps of I into $\underline{R}$ and for each i in $\underline{n}$

$$(c^*\nabla)_{\bar{1}}\ (e_i \circ c)=\sum_{j\in\underline{n}} h_i^j(e_j \circ c)$$

where for each j in $\underline{n}$. h_i^j is a smooth map of I into $\underline{R}$, Again by (12.9)

$$(c*\nabla)_{\bar{1}}g=\sum_{i\in\underline{n}} Dg_i(e_i \circ c)+\sum_{(i,j)\in\underline{n}\times\underline{n}} g_i h_i^j(e_j\circ c)$$

and so

$$(c*\nabla)_{\bar{1}}((c*\nabla)_{\bar{1}}g)=\sum_{i\in\underline{n}} D^2 g_i(e_i\circ c)+2\sum_{i\in\underline{n}}\Big(\sum_{j\in\underline{n}} h_j^i Dg_j\Big)(e_i\circ c)$$

$$+\sum_{i\in\underline{n}}\sum_{(j,k)\in\underline{n}\times\underline{n}}(g_j h_j^k h_k^i)(e_i\circ c)+\sum_{i\in\underline{n}}\Big(\sum_{j\in\underline{n}} g_j\, D h_j^i\Big)(e_i\circ c)$$

Hence $(c*\nabla)_{\bar{1}}((c*\nabla)_{\bar{1}}g)=0$ iff for each i in $\underline{n}$

$$D^2 g_i+2\sum_{j\in\underline{n}} h_j^i Dg_j+\sum_{(j,k)\in\underline{n}\times\underline{n}}(g_j h_j^k h_k^i)+\sum_{j\in\underline{n}} g_j\, D h_j^i=0$$

The above system of n equations of the second order can be tranformed into a system of $2n$ equation of the first order by putting $g_{n+i}=Dg_i$ moreover $g(t)=v$ and $((c^*\nabla)_1 g)(t)=w$ means that for each i in $\underline{n}$

$$g_{i(t)}=(d_{c(t)}\theta^i)(v)$$

and

$$\sum_{j\in\underline{n}} g_j(t)h_j^i(t)+Dg_i(t)=(d_{c(t)}\theta^i)(w)$$

so that

$$g_{n+i}(t) = (d_{c(t)}\theta^i)(w) - \sum_{j \in \underline{n}} g_j(t) h^i_j(t)$$

by ((5.12)...) the system of $2n$ differential equations of the first order with the initial conditions, given above, has a unique system of solutions, $g_1, ..., g_{2n}$ with domains I. Consequently there exists a unique map g of I into TX satisfying the conditions

$$(c^* \nabla)_{\bar{1}} \ ((c^* \nabla)_l \, g) + f = 0, \; g(t) = 0, \; ((c^* \nabla)_l \, g)(t) = w.$$

(14.11) Let X be a smooth manifold with covariant derivative ∇. For any two members u and v of $\Gamma^\infty(X)$. we write $T(u, v) = \nabla_u v - \nabla_v u - [u, v]$ and call the functions

$$\Gamma^\infty(X) \times \Gamma^\infty(X) \ni (u, v) \mapsto T(u, v) \in \Gamma^\infty(X)$$

the torsion field of ∇.

For any three members u, v, w of $\Gamma^\infty(X)$ we write

$$R(u, v)\, w = \nabla_u (\nabla_v w) - \nabla_v (\nabla_u w) - \nabla_{[u, v]} w$$

and call the function

$$\Gamma^\infty(X) \times \Gamma^\infty(X) \times \Gamma^\infty(X) \ni (u, v, w) \mapsto R(u, v)\, w \in \Gamma^\infty(X)$$

the curvature field of ∇.

From (C. D. 1) to (C. D. 4) it follows that T is bilinear and R is trilinear, and for all u, v, w in $\Gamma^\infty(X)$

$$T(u, v) = - T(v, u)$$

and

$$R(u, v)\, w = - R(v, u)\, w.$$

Now suppose that X is modelled on A^n. Let m be a smooth morphism of a smooth manifold Z modelled on A^p into X. Then for all smooth vector fields u and v on Z and all smooth liftings w of m we define

$$T((Tm) \circ u, (Tm) \circ v) = (m^* \nabla)_u (Tm) \circ v - (m^* \nabla)_v (Tm) \circ u - (Tm) \circ [u, v]$$

and

$$R((Tm) \circ u, (Tm) \circ v)\, w = (m^* \nabla)_u ((m^* \nabla)_v w) - (m^* \nabla)_v ((m^* \nabla)_u w) - ((m^* \nabla)_{[u, v]} w)$$

(14.12) Let X be a smooth manifold modelled on A^n, Let θ be a chart for X, let $g_1,\ldots,g_n$ be the n basic vector fields relative to θ and let ∇ be a covariant differentiation in X, if $T = 0$ then by (9.19) for all (i, j) in $\underline{n} \times \underline{n}$

$$\nabla_{g_i} g_j - \nabla_{g_j} g_i = [g_i, g_j] = 0$$

and so by (14.2) for all (i, j, k) in $\underline{n} \times \underline{n} \times \underline{n}$

$$\Gamma^i_{jk} = \Gamma^i_{kj}.$$

For all (j, k) in $\underline{n} \times \underline{n}$ put

$$T(g_j, g_k) = \sum_{i \in \underline{n}} T^i_{jk} g_j$$

where for all i in $\underline{n}$, T^i_{jk} is a smooth function on $Dm(\theta)$ into A. Then for all (i, j, k) in $\underline{n} \times \underline{n} \times \underline{n}$

$$T^i_{jk} = \Gamma^i_{jk} - \Gamma^i_{kj}$$

For all (i, j, k) in $\underline{n} \times \underline{n} \times \underline{n}$ put

$$R(g_i, g_j)\, g_k = \sum_{h \in \underline{n}} R^h_{ijk} g_h$$

where for all h in $\underline{n}$, R^h_{ijk} is a smooth function on $Dm\theta$ into A. Then for all (h, i, j, k) in $\underline{n} \times \underline{n} \times \underline{n} \times \underline{n}$

$$R^h_{ijk} = L_{g_i}(\Gamma^h_{jk}) - L_{g_j}(\Gamma^h_{ik}) + \sum_{m \in \underline{n}} (\Gamma^m_{jk}\Gamma^h_{im} - \Gamma^m_{ik}\Gamma^h_{jm})$$

(14.13) Let X be a smooth manifold modelled on A^n with the property that there exists a smooth vector fields $g_1,\ldots,g_n$ on X such that for each vector field v on X there exists n smooth functions $a_1,\ldots,a_n$ on X into A such that $v = a_1 g_1 + \ldots + a_n g_n$.

Let ∇ be the function on $\Gamma^\infty(X) \times \Gamma^\infty(X)$ into $\Gamma^\infty(X)$ such that for all vector fields

$$v = \sum_{i \in \underline{n}} a_i g_i \,, \qquad w = \sum_{i \in \underline{n}} b_i g_i$$

in $\Gamma^\infty(X)$ and for all x in X

$$(\nabla_v w)(x) = \sum_{j \in \underline{n}} (d_x b_j)(v(x)) g_j(x)$$

$$= \sum_{(i,j) \in \underline{n} \times \underline{n}} a_i(x)(d_x b_j)(g_i(x))g_j(x)$$

Then ∇ is a covariant differentiation in X, said to be induced by $g_1,...,g_n$.

For all i in $\underline{n}$

$$\nabla_v g_i = 0$$

and so for all (i, j, k) in $\underline{n} \times \underline{n} \times \underline{n}$ the components Γ^i_{jk} of ∇ are all zero.

We note that for all (i, j) in $\underline{n} \times \underline{n}$

$$T(g_i, g_j) = -[g_i, g_j]$$

By (14.9) for u', v', w' in $\Gamma^\infty(X)$

$$R(u', v')w' = 0$$

Now suppose that X is an open subset of $\underline{R}^n$. Let $(e_i; i \in \underline{n})$ be the canonical basis of $\underline{R}^n$. For each i in $\underline{n}$. Let D_i be the vector field on X such that for all x in X $D_i(x) = e_i$, Then $(D_i; i \in \underline{n})$ is a basis of $\Gamma^\infty(X)$ called canonical. In this case for all x in X

$$(\nabla_v w)(x) = \sum_{j \in \underline{n}} (Db_j(x) \cdot v(x))D_j(x)$$

$$= \sum_{(i,j) \in \underline{n} \times \underline{n}} a_i(x)Db_j(x) \cdot D_i(x) \cdot D_j(x)$$

Linear connection

(14.14) Let X be a C^∞ manifold modelled on a Banach space R over A. Let (η, Y, X) be a C^∞ vector bundle over X. By (8.4) $(T(\eta), T(Y), T(X))$ is a C^∞ vector bundle over $T(X)$ and since $\tau_X \, o \, T\eta = \eta \circ \tau_Y$ so $(\tau_X \circ T(\eta), T(Y), X)$ is a bundle over X. Form the direct sum $(\tau_X \oplus \eta, T(X) \oplus Y, X)$ of the vector bundles $(\tau_X, T(X), X)$ and (η, Y, X). Let c be a C^∞ morphism of the manifold $T(X) \oplus Y$ into the manifold $T(Y)$

Then a C^∞ manifold morphism of $T(X) \oplus Y$ into $T(Y)$ is called a linear connection in η iff the conditions: (C N. 1), (C N. 2), (C N. 3) and (C N. 4) are satisfied.

$$\text{(C N. 1)} \quad \tau_X \oplus \eta = \tau_X \circ (T\eta) \circ c$$

since $\tau_X \circ (T\eta) = \eta \circ \tau_Y$ for each x in X each v in $T_x X$ and each y in $\eta^{-1}[x]$, $(T\eta) \circ c(v, y)$ is in $T_x X$ and $\tau_Y \circ c(v, y)$ is in $\eta^{-1}[x]$.

(C N. 2) for each x in X, each v in $T_x X$ and each y in $\eta^{-1}[x]$;

$$(T\eta)\circ c(v, y) = v \text{ and } \tau_Y \circ C(v, y) = y$$

then for each x in X, each v in $T_x X$ and each y in $\eta^{-1}[x]$,

$$c(v, y) \text{ is in both } (T\eta)^{-1}[v] \text{ and } T_y Y.$$

(C N. 3) for each x in X and each y in $\eta^{-1}[x]$ the map

$$T_x X \ni v \mapsto c(v, y) \in T_y Y$$

is continuous and linear.

(C N. 4) for each x in X and each v in $T_x X$ the map

$$\eta^{-1}[x] \ni y \mapsto c(v, y) \in (T\eta)^{-1}[v]$$

is continuous and linear.

By abuse of language a linear connection in the tangent bundle τ_X of X is called a linear connection in X itself. Thus a linear connection in X is a C^∞ manifold morphism of $T(X)\oplus T(X)$ into $T(T(X))$, satisfing the conditions (C N. 1), (C N. 2), (C N. 3) and (C N. 4).

(14.15) Let U be an open subset of a Banach space R over A. Let S be another Banach space over A, and let $(pr_1, U\times S, U)$ be a trivial bundle over U. We identify $T(U\times S)$ with $(U\times S)\times(R\times S)$ and $T(U)\oplus(U\times S)$ with $U\times R \times U\times S$. Let c be a linear connection in the vector bundle $(pr_1, U\times S, U)$. Then c is a C^∞ morphism of $U\times R \times U\times S$ into $(U\times S)\times(R\times S)$ of the form

$$((u, r),(u, s)) \mapsto ((u, s),(r, -B(u, r, s))).$$

where $((u, r),(u, s)) \mapsto B(u, r, s)$ is a C^∞ morphism of $U\times R\times U \times S$ into S such that for each u in U, $(r, s) \mapsto B(u, r, s)$ is a bilinear map of $R\times S$ into S.

On the other hand let B be a C^∞ morphism of $U\times R\times U\times S$ into S such that for each u in U, $(r, s) \mapsto B(u, r, s)$ is a bilinear map of $R\times S$ into S. Then tha map c

$$U\times R\times U\times S \ni ((u, r),(u, s)) \mapsto ((u, s), (r, -B(u, r, s))) \in (U\times S)\times(R\times S).$$

is a linear connection in $(pr_1, U\times S, U)$.

(14.16) Let X be a C^∞ manifold modelled on R, let (η, Y, X) be a

C^∞ vector bundle over X with a framing of fiber type S and let c be a linear connection in η. Let (V, θ, ψ) be a frame for η, put $U = \theta[V]$. Then $(\theta \times id_s) \circ \psi$ is a C^∞ isomorphism of the manifold $\eta^{-1}[V]$ onto the manifold $U \times S$ and so $T((\theta \times id_s) \circ \psi)$ is an isomorphism of the tangent bundle $T(\eta^{-1}[V])$ onto the tangent bundle $T(U \times S)$ which is identified with $(U \times S) \times (R \times S)$. The map

$$((u, r),(u, s)) \mapsto (T((\theta \times id_s) \circ \psi)\,(c((T\theta)^{-1}(u, r),\ \psi^{-1}(\theta^{-1}(u), s)))$$

of $(U \times R \times U \times S)$ into $(U \times S) \times (R \times S)$ is a linear connection in the trivial vector bundle $(pr_1, U \times S, U)$, which is called the local expression of c relative to the frame (V, θ, ψ) from the first paragraph it follows that the local expression of c relative to (V, θ, ψ) is a map of the form

$$((u, r),(u, s)) \mapsto ((u, s),(r, -B(u, r, s))),$$

where $((u, r),(u, s)) \mapsto B(u, r, s)$ is a C^∞ morphism of $U \times R \times U \times S$ into S such that for each u in U,

$$(r, s) \mapsto B(u, r, s) \text{ is a bilinear map of } R \times S \text{ into } S.$$

For each point $((u, r),(u, s))$ in $(U \times R) \times (U \times S)$

$$c((T\theta)^{-1}(u, r),\ \psi^{-1}(\theta^{-1}(u), s)) = (\mathrm{T}\,(\psi^{-1} \circ (\theta^{-1} \circ \mathrm{id}_s)))\,((u, s),(r, -B(u, r, s)))$$

If $R = A^n$ and $S = A^p$ then for each u in U for each $r = (r_j; j \in \underline{n})$ in A^n and each $s = (s_k; k \in \underline{p})$ in A^p

$$B(u,r,s) = \left(\sum_{(j,k) \in \underline{n} \times \underline{p}} B^i_{jk}(u) r_j s_k ; i \in \underline{p} \right)$$

where for each i in $\underline{p}$, each j in $\underline{n}$ and each k in $\underline{p}$ the map $u \mapsto B^i_{jk}(u)$ is a C^∞ morphism of U into A, called a component of the linear connection c relative to the frame (V, θ, ψ) and denoted by C^i_{jk} by abuse of notation.

(14.17) Let X be a C^∞ manifold modelled on a Banach space R over A, let (η, Y, X) be a C^∞ vector bundle over X and let c be a linear connection in η. Let f be a C^k morphism of a C^∞ manifold Z modelled on a Banach space Q over A into Y. Let z be any point of Z and let x be the any point of X such that $f(z)$ is in $\eta^{-1}[x]$. Let w be any vector in $T_z Z$. Then $(T_x f)(w)$ is in $T_{f(z)}(\eta^{-1}[x])$. Since $\eta \circ f$ is a morphism of Z into X, $(T_z(\eta \circ f))(w)$ is in $T_x X$, therefore by (C N. 2). $c((T_z(\eta \circ f))(w),\ f(z))$ is in $T_{f(z)}(\eta^{-1}[x])$, and so

$$(T_z f(w)) + c((T_z(\eta \circ f))(w),\ f(z)) \text{ is in } T_{f(z)}(\eta^{-1}[x]) \text{ and so}$$

$$(T_z f(w)) + c((T_z(\eta \circ f))(w),\ f(z)) \text{ is in } T_{f(z)}(\eta^{-1}[x]).$$

Hence $(id_{\eta^{-1}[x]})_{f(z)}((T_z f)(w)) + c((T_z(\eta \circ f))(w),\ f(z))$ is a vector in $\eta^{-1}[\eta(f(z))]$ and is denoted by

$$C^z_w(f).$$

Now suppose that X and Z are open subsets of R and Q respectively. Let η be the trivial bundle $(pr_1, X \times S, X)$, where S is a Banach space over A. Then by (14.3) the linear connection c in pr_1 is a map of $X \times R \times X \times S$ into $(X \times S) \times (R \times S)$ of the form $((x, r), (x, s)) \mapsto ((x, s), (r, -B(x, r, s)))$where the map $B : ((x, r), (x, s)) \mapsto B(x, r, s)$ is a bilinear map of $X \times R \times X \times S$ into S for all x in X. Moreover the map f of Z into $X \times S$ is of the form $z \mapsto (f_1(z), f_2(z))$, where $z \mapsto f_1(z)$ and $z \mapsto f_2(z)$ are C^k maps of Z into X and S respectively. Let z be a point of Z, then $T_z Z = \{z\} \times Q$. Let q be any vector in Q, then

$$(T_z f)(z, q) = ((f_1(z), f_2(z)), (Df_1(z).q, Df_2(z).q))$$

and

$$(T_z(pr_1 \circ f))(z, q) = (T_z f_1)(z, q) = (f_1(z),\ Df_1(z)(q))$$

Hence

$$\begin{aligned} c((f_1(z),\ Df_1(z).q), (f_1(z), f_2(z))) \\ = ((f_1(z), f_2(z)), (Df_1(z).q, -B(f_1(z), \\ Df_1(z)(q), f_2(z)))) \end{aligned}$$

and so

$$C^z_{(z,q)}(f) = (f_1(z),\ Df_2(z).q + B(f_1(z),\ Df_1(z).q,\ f_2(z))$$

(14.18) Let m be a C^k morphism of a C^∞ manifold Z into a C^∞ manifold X, and let c be a linear connection in a C^∞ vector bundle (η, Y, X) over X. Then for each z in Z the map C^z

$$(T_z Z) \times Lift^k(m) \ni (w, f) \mapsto C^z_w(f) \in \eta^{-1}[m(z)]$$

is a covariant differntiation at z relative to m, said to be induced by the linear connection c.

In fact, C^z obviously satisfies (C D. 1), (C D. 2) and (C D. 3), to varify (C. D. 4) we may suppose that Z and X are open subset of Banach space over A and that η is a trivial vector bundle over X and then apply (14.17).

The covariant differentiation relative to m

$$\Gamma^k(Z) \times Lift^k(m) \ni (w, f) \mapsto (z \mapsto C^z_{w(z)}(f)) \in Lift^k_\eta(m)$$

is said to be induced by the linear connection c and is denoted by ∇ or simply by ∇.

(14.19) Let m be a C^k morphism of a C^∞ manifold Z into a C^∞ manifold X, let (η, Y, X) be a C^∞ vector bundle over X. let for each z in Z the map ∇^z be a covariant differentiaion at z relative to m, and let ∇ be the covariant differentiation relative to m such that for each z in Z, each w in $\Gamma^k(Z)$ and each f in $Lift^k_\eta(m)$,

$$(\nabla_w f)(z) = \nabla^z_{w(z)}(f)$$

Let z be a by point of Z and let f and g be any two members of $Lift^k_\eta(m)$ such that f and g coincide on an open neighbourhood W of z.

Now suppose that Z is finite dimensional , then there is a C^∞ map h of Z into A and an open neighbourhood V of z contained in W such that $h(u) = 0$ for all u in V and $h(u) = 1$ for all u in $Z - W$.

Since f and g coincide on W, $h(f - g)$ and $f - g$ coincide on Z. Hence for all vectors w in $T_z Z$.

$$\begin{aligned}\nabla^z_w(f) - \nabla^z_w(g) &= \nabla^z_w(f - g)\\ &= \nabla^z_w(h(f - g))\\ &= h(z)\nabla^z_w(f - g) + ((d_z h)(w))(f(z) - g(z)) = 0\end{aligned}$$

and so

$$\nabla^z_w(f) = \nabla^z_w(g).$$

(14.20) Let ∇ be a covariant differentiation in a finite dimensional C^∞ manifold *X*. then for each *x* in *X* and each pair of vector field *v* and *w* on *X*

$$(\nabla_w v)(x) = \nabla^x_{w(x)}(v)$$

let *m* be a C^k morphism of a finite dimensional C^∞ manifold *Z* into X. Let *z* be any point of *Z* and let *U* be any open neighbourhood of *m* (*z*) with the following properties, there exists n vector fields $v_1,...,v_n$ on *U* such that for each vector field *v* on *X* their exists *n* C^∞ morphism $h_1,...,h_n$ of *U* into *A* such that $m^{-1}[u] = Z$ and

$$v\restriction U = h_1 v_1 + ... + h_n v_n.$$

By (14.1) for each C^k vector field *v* on X, $v \circ m$ is a lifting of *m* to τ_X, but every lifting of *m* to τ_X is not of the form $v \circ m$. However, it is clear that every lifting of *m* to τ_X is a linear combination of $v_1 \circ m,...,v_n \circ m$ with coefficients from the ring of all C^k morphism of *U* into *A*. In view of the first paragraph for each C^k vector field *w* on *Z* and each member *f* of *Lift* $^k(m)$ we define $(m*\nabla)^z_{w(z)}(f)$ such that for each vector field *v* on *X*

$$(m*\nabla)^z_{w(z)}(v \circ m) = \nabla^{m(z)}_{Tm(w(z))}(v)$$

which apparently depends on $v_1,...,v_n$, but not in fact as we now show , let $u_1,...,u_n$ be *n* vector fields on *U* such that the restriction of each vector field *v* on *X* to *U* is a linear combination of $u_1,...,u_n$ with coefficients from the ring of all C^k morphism of *U* into *A*. Then for each *i* in $\underline{n}$

$$v_i = \sum_{j \in \underline{n}} h_{ij} u_j$$

where for each *j* in $\underline{n}$ h_{ij} is a C^k morphism of *U* into *A*. Let *f* be any member of *Lift* $^k(m)$ such that

$$f = \sum_{i \in \underline{n}} (h_i v_i) \circ m$$

Then

$$f = \sum_{i \in \underline{n}} (h_i (\sum_{j \in \underline{n}} h_{ij} u_j)) \circ m$$

$$= \sum_{(i,j) \in \underline{n} \times \underline{n}} (h_i h_{ij} u_j) \circ m$$

Now to prove to (C D. 1) to (C D. 4) for each vector field w on Z we have

$$\sum_{(i,j) \in \underline{n} \times \underline{n}} \nabla^{m(z)}_{(Tm)(w(z))} (h_i h_{ij} u_j)$$

$$= \sum_{(i,j) \in \underline{n} \times \underline{n}} (h_i(m(z)) h_{ij}(m(z))) \nabla^{m(z)}_{(Tm)(w(z))} (u_j)$$

$$+ \sum_{(i,j) \in \underline{n} \times \underline{n}} (d_{m(z)} h_i) h_{ij}(m(z)) u_j(m(z))$$

$$+ \sum_{(i,j) \in \underline{n} \times \underline{n}} (h_i(m(z)) (d_{m(z)} h_{ij})) u_j(m(z))$$

$$= \sum_{i \in \underline{n}} (h_i(m(z)) (\sum_{j \in \underline{n}} (d_{m(z)} h_{ij}) u_j(m(z)))$$

$$+ \sum_{i \in \underline{n}} (d_{m(z)} h_i) (\sum_{j \in \underline{n}} h_{ij}(m(z)) u_j(m(z)))$$

$$+ \sum_{i \in \underline{n}} (h_i(m(z)) (\sum_{j \in \underline{n}} (h_{ij}(m(z))) \nabla^{m(z)}_{(Tm)(w(z))} (u_j))$$

$$= \sum_{i \in \underline{n}} (h_i(m(z)) (\sum_{j \in \underline{n}} \nabla^{m(z)}_{(Tm)(w(z))} (h_{ij} u_j))$$

$$+ \sum_{i \in \underline{n}} (d_{m(z)} h_i) (\sum_{j \in \underline{n}} h_{ij}(m(z)) u_j(m(z)))$$

$$= \sum_{i \in \underline{n}} (h_i(m(z)) \nabla^{m(z)}_{(Tm)(w(z))} (v_i)$$

$$+ \sum_{i \in \underline{n}} (d_{m(z)} h_i) v_i(m(z))$$

$$= \sum_{i \in \underline{n}} \nabla^{m(z)}_{(Tm)(w(z))}(h_i v_i)$$

Hence for each z in Z, each vector field w on Z and each member f of $Lift^k(m)$ such that

$$f = \sum_{i \in \underline{n}} (h_i v_i) \circ m$$

we unambiguously define

$$(m * \nabla)^{-}_{w(z)}(f) = \sum_{i \in \underline{n}} \nabla^{m(z)}_{(Tm)(w(z))}(v_i)$$

and observe that $(m^* \nabla)$ is a covariant differentiation relative to m, assigning to each member of

$\Gamma^k(X) \times U$ a member of $Lift^k(m)$, and called tha pull back of through m.

(14.21) Let ∇ be a covariant differentiation in a C^∞ manifold X of finite dimension and let m be a C^k morphism of X into a C^∞ manifold Z of finite dimension. Let v and w be members of $\Gamma^k(X)$ and $\Gamma^k(Z)$ respectively such that $(Tm) \circ w = v \circ m$. Then for each members u of $\Gamma^k(X)$ and each z in Z

$$(m * \nabla)^{-}_{w(z)}(u \circ m) = \nabla^{m(z)}_{(Tm)(w(z))}(u)$$

$$= \nabla^{m(z)}_{v \circ m(z)}(u)$$

$$= (\nabla_v u) \circ m(z)$$

15

Geodesics

The notion of a Geodesic as curve with zero acceleration is the fundamental concept of Riemannian geometry, Geodesic minimizes arc length for points "sufficiently close". The study of tangent bundle of a differentiable manifold reduce the local study of geodesic on a differentiable manifold to the study of the trajectories of a vector field on tangent bundle. The introduction of the exponential map of an open set in tangent bundle to manifold is simply to collect all of the geodesies of manifold into a unique differentiable mapping. This notation is useful and permits to apply the inverse function theorem to show that any points of a differentiable manifold possesses a neighbourhood such that any two points of the neighbourhood can be joined by a unique geodesic which minimizes arc length.

(15.1) Let X be a finite dimensional C^{∞} manifold with covariant differentiation ∇ and let c be a C^k morphism of an open interval J or $\underline{R}$ into X. Then an element f of $Lift^k(c)$ is said to be parallel to c relative to ∇ iff

$$(c*\nabla)_{\bar{I}}(f) = 0$$

c is said to be autoparallel or a geodesic in X iff $(Tc)\circ\bar{1}$ which is an element of $Lift^k(c)$ is parallel to c.

i.e., for all t in J, $(c*\nabla)^{t}_{(t,1)}(Tc\circ\bar{1}) = 0$

(15.4) Let X be a C^∞ manifold modelled on $\underline{R}^n$, with covariant differentiation ∇; Let v be a C^k vector field on X such that $\nabla_v v = 0$, and let c be an integral solution for v at x in X. Then c is a geodesic in X, passing through x.

Proof:

$$\begin{aligned}(c*\nabla)_{\bar{1}}((Tc)\circ\bar{1}) &= (c*\nabla)_{\bar{1}}(v\circ c)\\ &= \nabla_{(Tc)\circ\bar{1}}(v)\\ &= \nabla_{v\circ c}(v)\\ &= (\nabla_v v)\circ c = 0\end{aligned}$$

Thus c is a geodesic in X passes through x.

(15.5) Let X be a C^∞ manifold modelled on $\underline{R}^n$ with covariant differentiation ∇, and let a C^k morphism c of an open interval J of $\underline{R}$ about O into X be a geodesic in X.

Then

$$(c*\nabla)_{\bar{1}}((Tc)\circ\bar{1}) = 0$$

Let (U,θ) be a chart at $c(0)$, Then by (14.7) for each t in J

$$(c*\nabla)^{t}_{\bar{1}(t)}((Tc)\circ\bar{1}) = \sum_{i\in\underline{n}} d_t(\underset{\sim}{\theta}^{n+i}\circ Tc\circ\bar{1})(\bar{1}(t))$$

$$+\sum_{(j,k)\in\underline{n}\times\underline{n}}(d_t(\theta^j\circ c))(\bar{1}(t))((d_{c(t)}\theta^k))(((Tc(\bar{1}(t))\Gamma^i_{jk}(c(t))e_i(c(t))$$

Hence by (8.2) for each i in $\underline{n}$

$$D(D(\theta^i\circ c)) + \sum_{(j,k)\in\underline{n}\times\underline{n}}(D(\theta^j\circ c))\,(D(\theta^k\circ c))\Gamma^i_{jk}\circ c = 0$$

called the classical differential equation of geodesies.

(15.6) Let X be a C^∞ manifold modelled on A^n with covariant differentiation ∇, then there exists a unique sprays s over X such that for each v in TX the map

$$K_\nabla(s(v)) = 0.$$

called the geodesic spray over X relative to ∇.

Proof: For each v in TX the map

$$T_v(TX) \ni z \mapsto ((T\tau_X)(z), K_\nabla(z)) \in T_{\tau_X(v)}(X) \times T_{\tau_X(v)}(X).$$

is a linear isomorphism, therfore there exists a unique elements (v) in $T_v(TX)$ such that $(T\tau_X)(s(v)) = v$ and $(K_\nabla(s(v))) = 0$ and so the maps $TX \ni v \mapsto s(v) \in T_v(TX)$ is evidently a second order differential equation over X. Next we show that s is a spray over X, so let a be any scalar in A and let h_a be the map of TX into itself such that for each x in X and each v in $T_x X$, $h_a(v) = av$. For each v in TX

$$\begin{aligned}(T\tau_X)(a(Th_a)(s(v))) &= a(T\tau_X)((Th_a)(s(v)))\\ &= a(T(\tau_X \circ h_a)(s(v)))\\ &= a(T\tau_X)(s(v)))\\ &= a\,v\\ &= (T\tau_X)(s(a\,v))\\ &= (T\tau_X)(s(h_a(v)))\end{aligned}$$

and

$$\begin{aligned}K_\nabla(a(Th_a)(s(v))) &= a\,K_\nabla((Th_a)(s(v)))\\ &= a\,h_a(K_\nabla(s(v)))\\ &= 0.\end{aligned}$$

Hence s is a spray over X. From the definition of s it follows that s is tha unique geodesic spray over X relative to ∇. We can show by using charts that s is C^∞.

(15.7) Let X be a finite-dimensional real C^∞- manifold with covariant differentiation ∇. If a C^k morphism g of an open interval J of $\underline{R}$ into X is a geodesic in X then $(Tg)\circ\bar{1}$ is an integral solution for the geodesic spray s over X relative to ∇. If a C^k morphism h of an open integral J of $\underline{R}$ into TX is an integral solution for s then c $= \tau_X \circ h$ is a geodesic in X, relative to ∇, and $\dot{c} = h$.

Proof: Let $g\colon J \to X$ be a geodesic in X. Then

$$K_\nabla \circ \left(T\left((Tg)\circ\bar{1}\right)\circ\bar{1}\right) = \left(g^*\nabla\right)_{\bar{1}}\left((Tg)\circ\bar{1}\right) = 0$$

Also $\tau_X \circ (Tg) \circ \bar{1} = g \circ \tau_J \circ \bar{1} = g$. Hence

$$T(\tau_X) \circ \left(T\left((Tg) \circ \bar{1}\right) \circ \bar{1}\right) = T\left(\tau_X \circ (Tg) \circ \bar{1}\right) \circ \bar{1} = (Tg) \circ \bar{1}$$

By(15.4)

$$K_\nabla \circ s \circ (Tg) \circ \bar{1} = 0 \text{ and } T\tau_X \circ s \circ (Tg) \circ \bar{1} = (Tg) \circ \bar{1}$$

Hence

$$\left(T\left((Tg) \circ \bar{1}\right) \circ \bar{1}\right) = s \circ (Tg) \circ \bar{1}$$

and so $(Tg) \circ \bar{1}$ is an integral solution for s.

On the otherhand, let h: $J \to TX$ be an integral solution for s.

Then $(Th) \circ \bar{1} = s \circ h$ and so

$$\begin{aligned}\left(T(\tau_X \circ h)\right) \circ \bar{1} &= T(\tau_X) \circ (Th) \circ \bar{1} \\ &= (T\tau_X) \circ s \circ h = h\end{aligned}$$

put $g = \tau_X \circ \dot{h}$, then

$$\begin{aligned}\left(g^*\nabla\right)_{\bar{1}}\left((Tg) \circ \bar{1}\right) &= \left(g^*\nabla\right)_{\bar{1}}\left((T\tau_X \circ h) \circ \bar{1}\right) = \left(g^*\nabla\right)_{\bar{1}} h \\ &= K_\nabla \circ \left((Th) \circ \bar{1}\right) \\ &= K_\nabla \circ s \circ h = 0\end{aligned}$$

Hence g is a geodesic in X relative to ∇.

Select Bibliography

1. Abraham, R. and J.E. Marsden (1978), Foundations of Mechanics, Second Edition, Benjamin-Cumming.
2. Abraham, R., J.E. Marsden, and T.S. Ratiu (1988), Manifolds, Tenser Analysis and Applications, Second Edition, Applied Mathematical Sciences 75, Springer-Verlag.
3. Arnold, V.I. (1989), Mathematical Methods of Classical Mechanics, Second Edition, Graduate, Texts in Mathematics 60, Springer-Verlag.
4. Artin, M. (2006), Algebra, Printice – Hall of India Private Limited New Delhi.
5. Bishop, R. and R. Crittenden (1964), Geometry of Manifolds, Academic Press.
6. Boothby, W.M. (1975). An Introduction to Differentiable Manifolds and Riemannian Geometry, Academic Press. New York.
7. Carmo, M.P. Do (1992), Riemannian, Geometry Birkhaiiser.
8. Cartan, H. (1970), Differential Forms, Hermann, Paris.
9. Chevalley, C.C. (1964), Theory of Lie Groups, Vol. I, Princeton University Press, Princeton, N.J.

10. Dieudonne, Jean (1960), Foundations of Modern Analysis, Academic Press, New York.

11. Ebin, D. (1970), The Manifold of Riemannian Metrics, Proc. Symp. Pure Math XV, AMS No. 26, pp. 11-40; and Bull Amer. Math Soc. No. 4, pp. 1002-1004.

12. Ebin, D. and J.E. Marsden (1970), Groups of Diffeomorphisms and the Motion of an in compressible Fluid, Ann. of Math 92, pp. 102-163.

13. Gorenstein, D. (1982), Finite Simple Groups, Plenum Press, New York.

14. Guillemin, V. and A. Pollack (1974), Differential Topology, Prentice Hall, Inc. Englewood Cliffs, N.J.

15. Helgason, S. (1978), Differential Geometry, Lie Groups and Symmetric Spaces, Academic Press.

16. Herstein, I.N. (1976), Topics in Algebra, 2nd ed., Wiley, New York.

17. Kelley, J.L. (1991), General Topology, Springer Verlag, New York.

18. Kobayashi and Nomizu (1996), Foundations of Differential Geometry, Vol. I, John Wiley & Sons, Inc.

19. Lang, S. (1965), Algebra, 2nd ed., Addison – Wesley, Reading, M.A.

20. Lang, S. (1993), Real and Functional Analysis, 3rd ed., Springer – Verlag.

21. Lang, S. (1996), Differential and Riemannian Manifolds 3rd ed., Springer – Verlag.

22. Marsden, J.E. (1974), Applications of Global Analysis to Mathematical Physics, Publish or Perish, Reprinted in "Berkeley Mathematics Department Lecture Note Series"

23. Miller, W. Jr. (1968), Lie Theory and Special Functions, Academic Press, New York.

24. Milnor, J. (1968), Differential Topology, Princeton University Press.

25. Munkers, J.R. (1975), Topology; A First Course, Prentice Hall, Englewood Cliffs, N.J.

26. Narasimhan, R. (1968), Analysis on Real and Complex Manifolds, North – Holland Publ., Co., Amsterdam.

27. Palasis, R.S. (1968), Foundation of Global Non-Linear Analysis. Benjanium, New York. Amsterdam.

28. Smale, S. (1967), Differentiable Dynamical System, Bull. Amer, Math Soc. 73, No. 6, pp. 747-817.

29. Sternberg, S. (1964), Lectures on Differential Geometry. Prentice – Hall.

30. Vladimirov, S.A. (1979), Groups of Symmetries of Differential Equations and Relativistic Fields, Atomizdat, Moscow.

31. Warner, F.W. (1983), Foundations of Differentiable Manifolds and Lie Groups, Springer, New York.

32. Wells, R. (1980), Differential Analysis on Complex Manifolds, Graduate Texts in Mathematics 65, Springer – Verleg.

33. Zharinor, V. (1992), Geometrical Aspects of Partial Differential Equations, World Scientific Singapore.

Index